Rules for Rebels

Rules for Rebels

The Science of Victory
in Militant History

Max Abrahms

OXFORD
UNIVERSITY PRESS

OXFORD

UNIVERSITY PRESS

Great Clarendon Street, Oxford, OX2 6DP,
United Kingdom

Oxford University Press is a department of the University of Oxford.
It furthers the University's objective of excellence in research, scholarship,
and education by publishing worldwide. Oxford is a registered trade mark of
Oxford University Press in the UK and in certain other countries

The moral rights of the author have been asserted

First Edition published in 2018
Impression: 4

Published in the United States of America by Oxford University Press
198 Madison Avenue, New York, NY 10016, United States of America

British Library Cataloguing in Publication Data

Data available

Library of Congress Control Number: 2018935624

ISBN 978–0–19–881155–8

Printed and bound in Great Britain by
Clays Ltd, Elcograf S.p.A.

To Karolina and other rebels

Table of Contents

List of Figures

List of Tables in Appendix

List of Tables in Appendix

Introduction: The Stupid Terrorist

Throughout history, militant groups have engaged in violence to overturn the political status quo. Some groups have battled to end poverty, persecution, and foreign occupation. Others have fought to achieve an independent state, communist revolution, or caliphate. What nearly all militants have in common is that their grievances exceed their capability to redress them. If their groups were stronger, militants wouldn't be fighting the government; with any luck, they would be leading it. The history of militant groups is thus a story about failure.

But not always. This book offers welcome news for the rebel. It analyzes hundreds of militant groups with a variety of social scientific methods from an array of disciplinary perspectives to discern the determinants of political success. The main take-away is that militant leaders possess a surprising amount of agency over their political destiny. Triumph is possible. It's neither arbitrarily nor structurally determined. But they need to know what to do. People often make the mistake of what's called "leader attribution error." They see something going right or wrong with a group and attribute the outcome to the skill of the leader.[1] This is the first book to identify a cohesive set of actions to enable militant leaders to win. Join me in exploring the secrets of their success. In the pages below, you'll discover the three simple rules of successful militant leaders. It turns out there's a science to victory in militant history. But even rebels must follow rules.

Leadership in Social Movements

Since the publication of Frederick Taylor's *Principles of Scientific Management* at the turn of the twentieth century, business schools and consultancy firms have studied companies in search of timeless leadership rules.[2] The very concept comes from business, with its longstanding belief in the power of leaders to lift a firm's performance.[3] Empirical studies from retail to education attest to the value of leadership on all sorts of outcomes.[4] A Harvard Business School study of forty-two industries shows that on average the effect of the CEO accounted for about 14 percent of the difference in corporate performance, ranging from about 2 percent in the meat processing industry to 21 percent in telecommunications.[5]

But my book is revolutionary. For starters, the field of international relations pays short shrift to the role of leaders.[6] As Daniel Byman and Ken Pollack have remarked, "Political scientists contend that individuals ultimately do not matter, or at least they count for little in the major events that shape international politics."[7] The political psychologist, Robert Jervis, observed several years ago, "The study of leadership has fallen out of favor in political science."[8] According to Michael Horowitz and Allan Stam, political scientists have consistently "ignored the role of leaders" over the past sixty years.[9] As Elizabeth Saunders notes, international relations theorists "have rarely incorporated a central role for leaders."[10] The limited work in international affairs focuses on specific individuals without formulating a generalizable theory of what smart leadership means or entails across time and space.[11]

Leadership neglect is particularly rampant in the study of international social movements. The sociologists Colin Barker, Alan Johnson, and Michael Lavalette acknowledge, "There is something of a black box in social movement studies in that leadership has been under-theorized."[12] The applied social psychologist, Bert Klandermans, agrees: "Leadership and decision-making aspects of social movement organizations...are more often debated than studied empirically...systematic studies of the way in which movement leaders function are scarce."[13] Sharon Erickson and Bob Clifford plead, "Whatever the reasons for scholars' relative neglect of the subject, we believe that leadership merits greater attention."[14] Aldon Morris and Suzanne Staggenborg also appeal that "social movement analysts need to open up the black box of leadership and develop theories and empirical investigations of how leadership affects the emergence, dynamics, and outcomes of social movements."[15] This book helps to fill that research lacuna.

In the early 1970s, Saul Alinsky published *Rules for Radicals* to help future community organizers wrest power from the political elite. In this primer for the "have-nots," the founder of modern community organizing shared targeted lessons he had learned over the years for successful protest.[16] But what about when protesters conclude that protesting isn't enough? Historically, many social movements have escalated to violence after nonviolence failed. In the late-nineteenth century, the leader of the Irish National Invincibles declared: "There comes an hour when protest no longer suffices. After philosophy, there must be action. The strong hand finishes what the idea has planned."[17] Michael Collins was convinced that "Irish Independence would never be attained by constitutional means" and that "when you're up against a bully you've got to kick him in the guts."[18] In his memoire, Irgun leader Menachem Begin described the Zionist group's predicament after nonviolence failed to protect the Yishuv: "What use was there in writing memoranda? What value in speeches? No, there was no other way. If we did not fight we should be destroyed."[19] When asked why they adopted violence in the 1950s, Algerian nationalists complained that the French had just shrugged off their futile strikes and boycotts.[20] In her autobiography of her time in the American Weather Underground, Susan Stern explained why the radical left-wing group escalated to violence: "As the years have passed, I've seen my efforts fail with thousands of others in the Civil Rights and anti-war movements. The time has come not merely to protest but to fight for what we believe in."[21] The leader of the Tamil Tigers shared a similar rationale for why his group embraced violence: "The Tamil people have been expressing their grievances . . . for more than three decades. Their voices went unheard like cries in the wilderness."[22] The African National Congress released a similar statement in July 1963 about why it ramped up its anti-apartheid tactics: "It can now truly be said that very little, if any, scope exists for the smashing of white supremacy other than by means of mass revolution action, the main content of which is armed resistance leading to victory by military means."[23] Accounts of the Syrian rebels suggest that many "picked up weapons as a last resort."[24] According to the Palestinian intellectual Zaid Abu-Amr, "As it became evident that the peace negotiations were not yielding any tangible results . . . Hamas was emboldened and became more aggressive in its opposition to the PLO and its tactics against Israel."[25] For this reason, some scholars expect violence whenever mass-based movements of nonviolent reform are politically unprofitable.[26] Like it or not, many radicals will become rebels. And there are rules for them, too.

The Rise and Fall of Islamic State

Islamic State (ISIS) may come to mind when you think of a savvy, successful militant group. Clad in black robes, ISIS leader Ibrahim Awad Ibrahim al-Badri al-Samarrai a.k.a. Abu Bakr al-Baghdadi ascended the pulpit of the Great al-Nuri mosque in the Iraqi city of Mosul on July 5, 2014 proclaiming the emergence of a new caliphate. In his Friday sermon, the self-proclaimed caliph announced to the *umma* that his foot-soldiers had just captured swathes of land in Iraq and Syria, effectively creating an Islamic State. "As for your mujahedeen brothers," he intoned, "Allah has bestowed upon them the grace of victory and conquest . . . He guided them and strengthened them to achieve this goal."[27] By year's end, ISIS would control one-third of Iraq and one-third of Syria—land mass roughly equal to the size of Great Britain— where the terrorists ruled over nine million people.[28] The Islamic State was bolstered by the largest influx of international jihadis in history. Over 40,000 foreign fighters from 110 countries headed to Syria and Iraq, more than four times the number of mujahedeen who had traveled to Afghanistan to battle the Red Army in the 1980s.[29] This recruitment rate was "unprecedented," as the head of the National Counterterrorism Center testified in early 2015.[30] ISIS' reach was hardly limited to the caliphate. With varying degrees of organizational involvement, scores of ISIS attacks in dozens of countries terrorized the world.[31] By 2016, Baghdadi had accepted the *bayat* or allegiance of forty-three terrorist group affiliates from Boko Haram in Nigeria to Abu Sayyaf in the Philippines. Not only did ISIS have territory and fighters, but cash. Billed as "the world's richest terrorist group," Islamic State raked in over a billion dollars a year from oil sales, taxes, looting, antiquity smuggling, and hostage-taking.[32]

The international media was quick to crown Islamic State leaders as terrorist masterminds. *The Guardian* credited its apparent feats to "highly intelligent leaders calling the shots."[33] In a story entitled "Military Skill and Terrorist Technique Fuel Success of ISIS," the *New York Times* gushed that the group's "battlefield successes" were due to the "pedigree of its leadership." The story concluded, "These guys know the terrorism business inside and out."[34] The *Financial Times* claimed that "ISIS is chillingly smart."[35] So, too, did the *Washington Post*, which described ISIS as "wildly successful" with its "calculated madness."[36] *Vox* likewise extolled ISIS as a "calculating, strategic organization."[37] *The Los Angeles Times* went even further, exalting ISIS leaders for having evidently "perfected their operations."[38] The word "sophisticated" was bandied about from *The Wall Street Journal* to *Foreign Policy* to characterize the "evil genius" of Baghdadi and his lieutenants.[39] If ever there was a smart, strategic militant group, Islamic State was apparently it.

This conventional wisdom in the media was fueled by think tank pundits, who said ISIS leaders were astute in three main ways.[40] First, ISIS leaders are smart to recognize the strategic utility of brutalizing civilians not only in its stronghold of Iraq and Syria, but in indiscriminate massacres throughout the world. In a *Politico* article entitled "How ISIS Out-Terrorized Bin Laden," Will McCants of the Brookings Institute explains: "The Islamic State's brutality . . . has been remarkably successful at recruiting fighters, capturing land, subduing its subjects, and creating a state. Why? Because violence and gore work."[41] The ISIS leadership may be immoral, but it's clever enough to appreciate that "brutality would be a winning political strategy" since "this terrifying approach to state building has an impressive track record."[42] McCants developed this argument into a major book, which stresses how ISIS' unmitigated savagery and gore are "the very qualities that made the Islamic State so successful."[43] Massacring the population is allegedly a better way to gain compliance than winning over its hearts and minds. As he puts it, ISIS "doesn't believe a hearts and minds strategy is effective, and for the past few years it has been proven right."[44] His Brookings hallmate, Shadi Hamid, shared this assessment with *National Public Radio* that indiscriminate "violence actually does work."[45] For Hamid, the late 2015 mass shooting at the Bataclan theater in Paris was a "smart move," as was blowing up the Russian passenger jet over the Sinai because such seemingly wanton acts of depravity are all part of the winning "method to ISIS's madness."[46] Hamid waxes eloquent about the strategic logic of killing civilians in his book: "Instilling terror in the hearts of your opponents undermines their morale, making them more likely to stand down, flee, or surrender" and "the willingness to inflict terrible violence has a deterrent effect, raising the costs for anyone who so much as thinks of challenging the group."[47] Echoing McCants, Hamid asserts that ISIS has proven that "capturing and holding large swathes of territory is possible" even "without the benefit of widespread popular sup-port."[48] In their bestseller on ISIS, Michael Weiss and Hassan Hassan repeat that "the group's notorious brutality helped it."[49] In countless media interviews and op-eds, they argue that the grisly head-chopping and cage-burning of hapless victims follows a "brutal logic" and indeed showcases "the genius of ISIS."[50] Clearly, pundits have been impressed with the ISIS leadership's strategy of sanctioning unbridled barbarism. Though sickened by the indiscriminate bloodshed like the rest of us, they claimed it nonetheless worked.

Second, pundits accredit the ISIS leadership's strategy of empowering opera-tives all over the world to maximize the bloodletting, largely by decentralizing the organization. Unlike more hierarchical groups such as Al Qaeda, which place a greater premium on educating, disciplining, and vetting members, the ISIS leadership takes a hands-off approach, beckoning fanatics across the globe to butcher civilians of their choosing in the group's name. According to Clint Watts of the Foreign Policy Research Institute, ISIS' achievements stem from the fact that the leaders have "empowered its networked foreign fighters to plan and perpetrate attacks at will."[51] The key to ISIS gains is that the leadership recognizes the benefits of "diffuse operational control," which grants extremists everywhere "the autonomy to plot and plan locally."[52] Fawaz Gerges likewise proclaimed that "the strategic logic of the Islamic State" is based on delegating tactical decision-making to extremists across the globe. This green light to slay anyone on their wish-list "enables ISIS to reap all of the benefits of an attack, while incurring none of the costs."[53] Peter Bergen of the New America Foundation also attributed the apparent success of the group to its diffuse structure. "What empowers ISIS," he wrote for *CNN.Com*, is that it "accepts all comers," encouraging risk-acceptant lunatics to travel to the caliphate, strike locally, or develop regional affiliates under the black banner.[54] "The brilliance of the ISIS system," expanded terrorism com-mentator Malcom Nance, "is that its recruitment system is almost passive." The self-described caliph invites every nutcase to the global massacre;

"Baghdadi welcomes them all."[55] The leaders could never have inflicted so much carnage on their own. But they were allegedly strategic enough to expand the bloodbath by decentralizing ISIS operations and recruitment.

Third, pundits laud the ISIS leadership's public relations strategy of broadcasting the group's misdeeds over social media platforms like Facebook, YouTube, and Twitter, thereby capturing the evil in graphic detail. ISIS has used social media to showcase many aspects of the group, but none more assiduously than its innovative sentencing techniques—from beheadings with a knife to decapitation through explosive detonation cord to death by dragging, drowning, immolation, burial, mashing, mutilation, stoning, roof-chucking, and squashing, sometimes with a tank.[56] Phillip Smyth of the Washington Institute for Near East Policy affirms that ISIS succeeded by developing "the perfect sociopathic image."[57] Colin Clarke of RAND and Charlie Winter of the International Centre for the Study of Radicalisation proclaim that the "quality" of ISIS propaganda "truly has been unmatched" in human history.[58] In a story called "How the Islamic State's Massive PR Campaign Secured its Rise," an anti-Russia think tank known as Bellingcat explained that their "Public relations programs are perhaps singlehandedly responsible for their success in both recruiting foreigners, and even seizing control of sizeable swaths of land in Iraq and Syria."[59] Weiss and Hassan add that ISIS' secret sauce has been its "slick propaganda machine," especially its "peerless ability to produce sleek, hour-long propaganda and recruitment films."[60] The Al Qaeda in Syria specialist Charles Lister of the Middle East Institute also attributes ISIS success to its "slick propaganda media releases," as these "jihadists in particular proved especially adept at managing their use of social media and the production of qualitatively superior video and imagery output."[61] These pundits are in good company with this narrative; a search for the terms "slick," "video," and "ISIS" on Google yielded over 5 million hits in November 2014 alone.[62] The international media has promoted this group-think. "Where the Islamic State innovated the most," the *New York Times* gushed, "was in carrying out increasingly gruesome violence explicitly to film it—to intimidate enemies and to draw recruits with eye-catching displays on social media...Those techniques have proved so effective."[63] *Business Insider* raved, "It's well-known that ISIS has been very successful" because of its gruesome online propaganda.[64] *Wired* magazine went so far as to say that "Islamic State has been singularly successful" because of its unique ability to "inspire dread" and "cultivate this kind of image."[65] Advertising grotesque atrocities against innocent people is apparently terrific for militant groups and ISIS does it best.

And yet, something unexpected happened. The ISIS caliphate died as quickly as it had appeared. Pundits had been too busy glorifying its strategy to realize the group was losing. Indeed, ISIS was failing by its own standards.

In 2015, ISIS territory shrank by 40 percent in Iraq and 20 percent in Syria.[66] ISIS lost another quarter of its territory the following year.[67] By early 2017, ISIS had ceded two-thirds of its land.[68] By springtime, ISIS controlled less than 7 percent of Iraq and was being defeated in Syria by the Syrian Arab Army, its Shia militia partners, American and Russia airpower, Kurdish warriors, and a smattering of other militants.[69] In his May 2017 Pentagon press conference, Defense Secretary James Mattis noted that "ISIS had lost 55,000 square kilometers and regained none of it."[70] The leading Arab daily conceded, "ISIS is battered and in retreat, and its alleged 'caliphate' is nearly destroyed on the ground."[71] *The Guardian* acknowledged later that summer "the crumbling of the ISIS caliphate," as "black flags are no longer flying."[72] Even its erstwhile capitals in Mosul and Raqqa were falling fast. Former Defense Secretary, Robert Gates, was prepared to call it then—ISIS had "lost" in its core goal of establishing a caliphate.[73] Tellingly, that June ISIS blew up the al-Nuri mosque—the very site where the caliphate had been declared.[74] Iraqi Prime Minister, Haider al-Abadi, described the ISIS own-goal as "an official announcement of their defeat."[75] Later that month, from the ruins of al-Nuri, the Iraqi military spokesman faced no ISIS opposition when he declared "their fictitious state has fallen."[76] Antony J. Blinken, the Deputy Secretary of State under Obama, noted: "Its core narrative—building an actual state—is in tatters."[77] As the group lost land, its revenues also shrank until it could no longer pay fighters, which spurred defections and dissuaded recruits from joining the losing team. Air Force Major General Peter Gersten captured the sorry state of Islamic State: "We're seeing a fracture in their morale; we're seeing their inability to pay; we're seeing the inability to fight; we're watching them try to leave Daesh in every single way," using an Arabic acronym for Islamic State.[78]

But nowhere was the ISIS collapse clearer than in its once-vaunted propaganda messaging. ISIS spokesman, Abu Mohammad al-Adnani, had gone from encouraging Muslims to perform *Hijrah* or emigration to the caliphate to telling jihadi-wannabees to stay home.[79] Adnani was a dead man by the summer of 2016, as would soon be the guy in charge of producing the ISIS snuff videos, Abu Mohammed al-Furqan, and then his replacement, Abu Bashir al-Maslawi. The content of the videos became almost laughable. ISIS was reduced to broadcasting pictures of a one-legged suicide bomber in need of a cane, while begging members of the apocalyptic group to stop scurrying away from the battlefield.[80] A Syrian opposition activist told the *Associated Press* in June 2017: "The propaganda of the organization has become zero to be frank. It indicates their collapse and that the group is retreating."[81] Even the sympathizers or so-called "fan-boys" in pro-ISIS chat rooms conceded the caliphate project was a complete bust.[82] As one journalist put it, "Defeat is hard to sell."[83] Baghdadi was reportedly seen looking "thin and stooped."[84]

Although ISIS' raison d'être of a caliphate went up in smoke, there was a clear winner—its arch enemies. The Salafi jihadists repeatedly said the Islamic State was intended to curb the influence of Iran and its Shia proxies, especially Hezbollah. But instead of becoming the seat of a hardline Sunni state, Iraq and Syria were turned into Shia-country.[85] The sociologist William Gamson observes, "There is no more ticklish issue . . . than deciding what constitutes success."[86] The political scientist, James DeNardo, remarks that we cannot ever be sure whether terrorists get what they want; at best, we can evaluate whether they got what they claimed to want.[87] The Islamic State project face-planted by this standard. They were "evil losers" as President Trump likes to say.

This reversal of fortune was actually quite predictable. From the moment Baghdadi declared a caliphate in 2014, I gave hundreds of media interviews from the *Associated Press* to the *BBC* pointing out the basic analytical problem— the very behaviors lauded as strategic have historically doomed militant groups; ISIS, I always said, would be no exception.[88] With a little historical context and methods training, it was obvious that Baghdadi was no master-mind and neither were his fellow strategists. They were, as you'll see, supremely stupid terrorists. President Barack Obama got blasted in the media for saying early on that ISIS was the "JV team" (i.e., Junior Varsity) of terrorists. Actually, he was right—at least when it came to their cluelessness about devising a winning long-term strategy.

The Rules for Rebels

Smart militant leaders do three simple things for victory:

1. They recognize that not all violence is equal for achieving their stated political goals. In fact, smart leaders grasp that some attacks should be carefully avoided because they hurt the cause. My research is the first to empirically show that there's variation in the political utility of militant group violence depending on the target. Compared to more selective violence against military and other government targets, indiscriminate violence against civilian targets lowers the likelihood of political success. So, the first thing smart militants do is recognize that civilian attacks are a recipe for political failure. You might say that the first rule for rebels is to not use terrorism at all. There's no consensus over the definition of terrorism.[89] But most scholars define it as attacks against civilian targets in particular.[90] As the terrorism scholar, Louise Richardson, remarks, "The defining characteristic of terrorism is the deliberate targeting of civilians."[91] The legal scholar Alan Dershowitz also notes, "The deliber-ate killing of innocent civilians is a central element in most definitions of

terrorism."[92] The economists Walter Enders and Todd Sandler agree, "Virtually all definitions consider terrorist attacks against civilians as terrorism."[93] When talking about terrorism, we mean attacks on civilian targets like schools, markets, movie theaters, mosques, rock concerts, soccer games, synagogues, commercial airplanes, cruise-ships, churches, businesses, and apartment buildings unless occupied by military personnel. We're not talking about blowing the treads off a tank. What matters for the rebel, though, isn't how we define terrorism, but that he learns the folly of harming civilians. In this book, that means opposing terrorism.

2. The second rule is to actively restrain lower-level members from committing it. It doesn't matter whether the rebel understands the futility of terrorism if his members continue to do it anyway. The key is for him to take a stand against terrorism and build the organization so lower-level members abide. Centralizing the organization is invaluable for educating fighters about which targets to avoid, disciplining wayward operatives for harming civilians, and vetting out members who seem prone to undermining the cause with terrorism.

3. And the third rule for rebels is to distance the organization from terrorism whenever subordinates flout their targeting guidelines by attacking civilians. Like CEOs, smart militant leaders know how to brand their organization for maximum appeal when members publicly shame it. In practice, this means disavowing terrorism in all sorts of scientifically proven ways to project a moderate image of the group even when members act otherwise. In sum, smart rebels learn that tactical moderation pays, restrain lower-level members so they comply, and mitigate the reputational costs even when they don't.

As you'll see, these rules for rebels are based on insights from numerous academic disciplines (e.g., communication, criminology, economics, history, management, marketing, political science, psychology, sociology) and methodological approaches (e.g., field research, qualitative cases studies, content analysis, network analysis, regression analysis, survey experiments). Although this book applies these lessons to hundreds of militant groups throughout the world, one group gets more attention than any other—Islamic State. ISIS gets more ink because of its intrinsic importance as a militant group. This is the worst group the world has ever known in terms of its civilian carnage, geographic reach, and global terror. Conveniently, these characteristics also mirror the rules for rebels, so it's ideal for illustrating to future generations what not to do.

Ironically, the very behaviors described as strategic did more to destroy the caliphate than create it. For starters, ISIS didn't acquire its stronghold from terrorism. Patrick Skinner, the former CIA counterterrorism case officer, points out that ISIS' territorial gains in Iraq and Syria "were not terrorist... successes."[94] As the historian Walter Laqueur explains, ISIS set up shop in areas "sparsely populated or not populated at all...these regions were a kind of no man's land in which government forces had lost control, but had not been replaced by any other authority." The so-called caliphate was "more like a power vacuum than a new state."[95] Even in densely populated Mosul, ISIS just stormed into the military void. Although the U.S. had supplied billions of dollars to the Iraqi military over the years, it was woefully unprepared—weapons systems had fallen into disrepair, the officer corps was swollen with political hacks, thousands of "ghost soldiers" short on ammo were on the books, but never even showed up to fight.[96] As Brian Fishman observes, "ISIS did not actually fight its way into Mosul...the army collapsed."[97] More to the point, Mosul residents admit they weren't terrorized or coerced into support-ing the group. On the contrary, they "initially welcomed the Islamic State" because they thought it would offer protection.[98] The support of locals dissi-pated, however, the moment the group terrorized them. This indiscriminate ruthlessness ensured ISIS was reviled in every town, ostracized even by other Salafist groups, grossly out-manned in every battle from Tikrit to Raqqa, and unable to hold onto its own fighters, who increasingly defected due largely to the wanton killing.[99] The more terrorism ISIS inflicted, the less territory it controlled.[100] The decentralization of ISIS' violence all over the world united it against the group. Made up of over sixty-five countries, the anti-ISIS coalition had no trouble gaining participants precisely because the depraved group threatened citizens everywhere, turning even erstwhile friends like Qatar, Saudi Arabia, and Turkey into enemies. This anti-ISIS coalition was neither automatic nor inevitable. The turning point came in August 2015, when ISIS made the fateful decision to behead the American journalist, James Foley, and then boast about it over social media. ISIS said in the video that the beheading was intended to deter the U.S. from going after the group.[101] ISIS leaders launched hundreds of Tweet-storms threatening to attack the U.S. if it inter-vened militarily against the fledgling caliphate.[102] Yet the video galvanized Obama to assemble the largest counterterrorism coalition ever assembled to pummel the group. ISIS paid a steep, albeit predictable price for breaching the rules for rebels. Baghdadi was a stupid terrorist notwithstanding the think tank consensus.

Book Layout

The three rules for rebels are developed sequentially in this book. In the first part, I develop the first rule for rebels. Specifically, I present evidence from a host of methodological approaches—field research, case studies, regression analysis, and experiments—that certain kinds of attacks are more effective than others. These chapters highlight that militant groups are far more likely to achieve their political demands when violence is directed against military targets rather than civilian ones. This original finding going back to my days in the West Bank during the Second Intifada flies in the face of what I call the Strategic Model of Terrorism.[103] As its name suggests, this school of thought, prominent in political science, assumes that groups turn to terrorism because it offers the best chance to attain their political goals. In fact, my research reveals that civilian attacks backfire by lowering the prospects of government concessions—not to mention organizational survival. Leaders may not initially grasp the risk of terrorism, but smart ones learn it over time. Learning to win by sparing civilians is the first and foremost rule for militant leaders. Without internalizing this rule, they can't be expected to follow the other ones and prevail.

The second part of the book spells out the second rule for rebels. In these chapters, I show how smart leaders prevent their fighters from harming civilians, boosting the likelihood of victory. Some political scientists, like Jeremy Weinstein and Stathis Kalyvas, agree that civilian attacks are self-defeating for militant groups at least in civil wars. But these scholars write out the role of the leader by attributing indiscriminate violence to structural conditions such as the availability of resources rather than to decisions from the top.[104] I restore agency to leaders by disclosing how smart ones maintain operational control under any structural condition, effectively restricting their members from sabotaging the cause with terrorism. The physical scientist, Brian Jackson, defines operational control as "The ability to control or influence the activities and operations being carried out in pursuit of the organization's strategic goals."[105] Restraining lower-level members of the organization is tricky because they're less likely to recognize the political costs of terrorism or even care about them. But leaders aren't passive observers of militant group behavior; rather, they wield considerable influence over the tactical choices of their subordinates and hence the likelihood of victory. Leaders can dramatically reduce the use of terrorism in their ranks simply by telling them not do it and then building the organization so they comply. As we'll see, a centralized group is critical for instilling tactical restraint by helping the leader to communicate which targets to avoid, punish targeting violations against civilians, and vet out wannabe-members inclined to commit them. Whereas the first part of the book shows that smart leaders recognize the value of civilian restraint, the second part shows how to achieve it by getting members to comply with it.

The third part of this book explains how smart leaders respond when they don't. Following the first two rules for rebels substantially reduces terrorism in their ranks. But it doesn't eliminate it entirely. Even leaders who follow both rules will occasionally face a public relations fiasco. Luckily for rebels, there's a considerable body of academic research on how organizations can restore their image when members engage in face-threatening behaviors that risk tarnishing it. Drawing from the fields of communication, marketing, and psychology, I show how smart militant leaders adopt the same proven crisis management strategies as CEOs to maximize the appeal of the organization when members publicly shame it. Branding is essential for all organizations, but especially militant groups whose fate depends on garnering international and local support. For the militant leader, the key to developing a winning brand is by distancing the organization from terrorism when operatives per- petrate it. In practice, when operatives kill civilians, this means engaging in scientifically-based denial strategies to demonstrate goodwill. These three rules for rebels—learning, restraining, and branding to win—are the secrets for victory. Long before ISIS inverted this playbook, successful militant leaders were following it.

Notes

1. Nye 2008, 3.
2. Taylor 1914.
3. Barker 2001, 66.
4. Jones and Olken 2005, 835; Hall et al. 1986, 5.
5. Wasserman et al. 2001.
6. For exceptions, see Barber 1992; Horowitz et al. 2015; Saunders 2011.
7. Byman and Pollack 2001, 108.
8. Jervis 2013, 154.
9. Horowitz et al. 2015, xii.
10. Saunders 2011, 3–4.
11. Horowitz and Stam 2014, 528.
12. Barker et al. 2001, 1.
13. Barker, Johnson, and Lavalette 2001, 3.
14. Nepstad and Clifford 2006, 1.
15. Morris and Staggenborg 2004, 190.
16. Alinsky 1971.
17. Tynan 1894, 488.
18. Boot 2013, 250.
19. Begin 1978, 43.
20. Sharp and Finkelstein 1973, 544.
21. Stern 2007, 130–1.
22. Richardson 2007, 50.
23. Turok 2003, 277.
24. Lister 2016, xiii.
25. Abū 'Amr 1994, 75.
26. Lichbach 1998, 59.

27. Baghdadi, 5 July, 2014.
28. Johnson 2016.
29. Callimachi, March 29, 2016; Fishman 2016.
30. Crawford and Koran, February 11, 2015.
31. Lister et al., February 13, 2017.
32. Heißner et al. 2017.
33. Chulov, September 7, 2016.
34. Hubbard and Schmitt, August 27, 2014.
35. Gardner, 18 February, 2015.
36. McCoy, June 13, 2014; McCoy August 12, 2014.
37. Fisher, November 23, 2015.
38. Williams, November 22, 2015.
39. See, for example, Andrews and Schwartz, 22 August 2014; Brannen, 21 August 2014.
40. To be clear, analysts touted Islamic State's behavior as strategic. In no way did they approve of the group's behavior. All of the analysts referenced in this book are on the same counterterrorism team as I am. My disagreement with them is not normative. It is strictly analytical in terms of which militant group behaviors facilitate political success.
41. McCants, August 19, 2015.
42. Ibid.
43. McCants 2015, 126.
44. Ibid. 151, 159.
45. "In Recruitment Efforts," November 16, 2014.
46. "Why U.S. Governors Shouldn't," November 17, 2015; Hamid, November 24, 2015.
47. Hamid 2016, 221.
48. Ibid. 232.
49. Weiss and Hassan 2015, 229.
50. Hassan, February 7, 2015.
51. Watts, November 23, 2015.
52. Watts, April 4, 2016.
53. Gerges August 14, 2016.
54. Bergen, November 5, 2015.
55. Nance 2016, 210–11.
56. Ibid. 256.
57. Gebeily, June 26, 2014.
58. Clarke and Winter, August 17, 2017.
59. Goldsmith, February 11, 2015.
60. Weiss and Hassan 2015, xv, 171.
61. Lister 2016, 240, 4.
62. Stern and Berger 2015, 120.
63. Barnard and MacFarquharnov, November 20, 2015.
64. Engel November 28, 2015.
65. "Why ISIS Is Winning the Social Media War," April, 2016.

66. Rosen, January 5, 2016.
67. "Islamic State Group 'Lost Quarter of Territory' in 2016," January 19, 2017.
68. DiChristopher, March 31, 2017.
69. "Islamic State Has Lost Most Territory it Held in Iraq," April 11, 2017.
70. Starr, May 24, 2017.
71. Ignatius, May 27, 2017.
72. Chulov, June 4, 2017; Chulov, September 7, 2016.
73. Kelly, May 23, 2017.
74. "Battle for Mosul," June 22, 2017.
75. Alkhshali et al., June 23, 2017.
76. McKernan, June 29, 2017.
77. Blinken, July 9, 2017.
78. "Fewer Foreign Fighters Joining Islamic State," April 26, 2016.
79. Mushtaq, September 6, 2017.
80. "ISIS Propaganda Machine," June 11, 2017.
81. Ibid.
82. Lichfield, June 9, 2017.
83. Bershidsky, June 11, 2017.
84. Chulov, June 4, 2017.
85. Abi-Habib, April 3, 2017.
86. Gamson 1975, 248.
87. DeNardo 1985, 63.
88. See, for example, Abrahms, 2 September 2014; Abrahms 1 September 2014; Abrahms, 26 September 2014; Abrahms, 26 August 2014; Abrahms, 19 August 2014; Abrahms, 4 August 201; Abrahms, 16 March 2015; Abrahms, 20 April 2015; Abrahms, 4 October 2014; Kavanaugh 24 June 2015.
89. Schmid and Jongman, 2005, 6, 28.
90. See, for example, Wilkinson 1986, 54, 56.
91. Richardson 2007, 6.
92. Dershowitz 2002, 4.
93. Enders and Sandler 2012, 3.
94. "IS Top Command Dominated by Ex-Officers," August 10, 2015.
95. Laqueur 2016, xix.
96. Cordesman and Khazai 2014.
97. Fishman 2016, 199.
98. MacDiarmid, July 28, 2017.
99. Freeman, June 21, 2014.
100. Bhojani, August 21, 2017.
101. Carter, August 20, 2014.
102. Stern and Berger 2015, 159.
103. Abrahms 2008.
104. See, for example, Weinstein 2006; Kalyvas 2006; McCarthy and Zald 1977.
105. Jackson 2006, 244.

Rule #1
Learning to Win

1

My West Bank Discovery

Shortly after the September 11, 2001, terrorist attacks, I headed to the West Bank to do field research. This was at the height of Palestinian terrorism during the so-called Second Intifada. And Israel was in the process of building up a massive security barrier to prevent suicide attackers from crossing into pre-1967 territory. Officially, I was doing a fellowship in the Moshe Dayan Center at Tel Aviv University. But I used the apartment of my friend from ABC News in Jerusalem as a safe-house to launch excursions. By day, I would take a cab into West Bank towns, asking Palestinians what they thought of the new wall. At night, I would return to Jerusalem and ask Israelis their opinion of the fence. The choice of words was deliberate. Palestinians call it a wall to emphasize the permanence of the Israeli occupation, whereas Israelis call it a fence to signal their preference for democratic norms over a police state. I wanted to understand the wall-fence from both perspectives and come up with a dissertation topic to begin the following year.

The Israelis confessed that they weren't proud of the fence. It was ugly, not just aesthetically but as a symbol of failure to peacefully resolve the conflict with their Palestinian neighbors. The fence, they taught me, wasn't a new idea. It had been bandied about for ages. What changed was that more Israelis had now given up hope that the Palestinians could ever be trusted as a partner for peace. Polls confirm that most Israelis initially favored the Declaration of Principles known as the Oslo peace process in the 1990s, but soured on the land-for-peace deal after a series of Palestinian terrorist attacks.[1] Beginning in the early 1990s, the Tami Steinmetz Center for Peace Research polled a representative sample of Israelis on two questions: Do you believe the Palestinians are potential partners for peace? And do you support the peace process? Instances of Palestinian terrorism systematically inclined Israelis to answer "no" to both questions.[2] Although some religious Jews always wanted to seize the West Bank for biblical reasons, the fence gained broad-based popularity in Israel as a unilateral counterterrorism measure in the apparent absence of a trustworthy bargaining partner. Understandably, the Palestinians I spoke to had a very different take on the wall. They told me how harmful the wall is for their pride, economy, health, even for visiting friends and family on the other side. But mainly they stressed the damage to their political aspirations. Every Palestinian I met seemed depressed; the wall shattered their dreams of a viable independent state.

One day, from the back of a cab in Ramallah, I began reading an article from the flagship journal in political science I had printed out the night before. In "The Strategic Logic of Suicide Terrorism," Robert Pape contends in his now-famous study in the *American Political Science Review* that terrorism is a "strategic decision by leaders of the terrorist organizations" who "have learned that it works" based on "quite reasonable assessments of the outcomes."[3] According to Pape, terrorism "pays" because of its ability to "cause mounting civilian costs to overwhelm the target state's interest in the issue in dispute and so to cause it to concede the terrorists' political demands."[4] Specifically, he maintains that "terrorism follows a strategic logic" based on its history of "success" in coercing "significant territorial concessions."[5] Pape points to "Israeli territorial concessions to the Palestinians" as his main evidence for the "coercive effectiveness of this strategy."[6] Although Palestinian suicide terrorism is his primary case study to show the coercive effectiveness of terrorism, Pape says 9/11 and other terrorist attacks follow the same strategic logic.

I was immediately struck by the disconnect between his claims and the reality on the ground. From Israelis, I discovered that Palestinian terrorism eroded support for territorial concessions while creating support for a fence. And from Palestinians, I heard their pain over how the wall blocked statehood. The Palestinians may have looked like political winners from the University of Chicago, but not from Ramallah. At least, none of the Palestinians with whom I had tea saw themselves as winners. I couldn't help but wonder: If the Palestinians are the big political success story, then how have other groups fared that also used terrorism? And if terrorism isn't an effective political tool, then why do so many groups attack civilians?

In graduate school, I set out to read everything on the political effects of terrorism. To my surprise, I quickly discovered there was almost no empirical research on whether terrorism actually helps groups to achieve their demands. Martha Crenshaw observed that "the outcomes of campaigns of terrorism have been largely ignored" as "most analyses have emphasized the causes and forms rather than the consequences of terrorism."[7] Ted Robert Gurr likewise noted that terrorism's political effectiveness is "a subject on which little national-level research has been done, systematically or otherwise."[8] This omission in the literature came as a shock. In addition to Robert Pape, other prominent political scientists like David Lake and Andrew Kydd and Barbara Walter also subscribe to what I call the Strategic Model of Terrorism.[9] As its name suggests, this school of thought posits that groups turn to terrorism because of its strategic value in pressuring government concessions.[10] Various literatures mention dozens of uses for violence.[11] But the Strategic Model is explicit that "an organization's success or failure is measured in terms of its ability to attain its stated political goals."[12] In the West Bank, I had found my dissertation topic. At Stanford University as a pre-doctoral and post-doctoral fellow, I began publishing the first systematic empirical studies to test it.

My focus has been on what I have dubbed as the outcome goals of militants rather than their process goals. Process goals are intended to sustain the group by securing financial support, attracting media attention, scuttling organization-threatening peace processes, or boosting membership and morale, often by provoking government overreaction. The outcome goals of militants, by contrast, are their stated political ends, such as the realization of a Palestinian homeland, the removal of foreign bases from Greece, or the establishment of Islamism in India. An important difference between process goals and outcome goals is that unlike the former, the latter can only be achieved with the compliance of the target government.[13] Focusing on outcome goals makes sense for several reasons. They're the key measure of success according to the Strategic Model. When political scientists say that terrorism works, they don't mean that it scares people or attracts media attention. If those were the measures for victory then terrorism, by definition, would have a 100 percent success rate. When the international media and think tanks said ISIS was a strategic success they pointed to its (short-lived) caliphate. As the terrorism historian Richard English notes, "Any full understanding of a phenomenon which—like terrorism—is focused on the pursuit of political change, will necessitate analysis of how far such change has actually been achieved."[14] In his book on violent protest, James DeNardo agrees: "The effectiveness of demonstrations depends ultimately on the responsiveness of the regime to disruption in the streets. Roughly speaking, we mean by responsiveness the regime's willingness to trade concessions for tranquility."[15] Militants admit their outcome goals are paramount. As one Provisional Irish Republican Army member put it: "The success of an undertaking must be judged against its reasons for beginning and its end, and, thus seen, the IRA failed utterly."[16] Another PIRA member elaborated: "To what extent did the IRA achieve their strategic goals? I would have to say they failed dismally . . . the IRA completely failed to achieve its strategic goal, which was to force a British declaration of intent to withdraw."[17] His buddy in the group reiterated: "The political objective of the Provisional IRA was to secure a British declaration on intent to withdraw. It failed."[18] As Seneca observed, "No one proceeds to shed human blood for its own sake, or at any rate few do so."[19] Even jihadists claim to value outcome goals. For all the talk about Islamist nihilism, "Jihadism conceives of jihad in strategic terms." Jihad expert Nelly Lahoud explains, "Despite the theological framework that governs the rhetoric for some jihadi ideologues . . . they want Islam to be instrumental to strategic/political change."[20] My West Bank experience made me question the conventional wisdom of whether terrorism is indeed effective for achieving outcome goals. But what about for other groups throughout history? Pape said his thesis applies to 9/11, so let's start there.

The Political Failure of 9/11

Osama Bin Laden said the 9/11 attacks were intended to redress four grievances that he had been railing against throughout the 1990s. First, Bin Laden's best-known ultimatum was for the U.S. to withdraw its troops from the Persian Gulf. What he found objectionable wasn't just the stationing of troops in the "Land of the two holy places" during the 1991 Gulf War, but that the bases in Saudi Arabia were then used as a "spearhead through which to fight the neighboring Muslim peoples."[21] Bin Laden's criticisms of U.S. military inter-ference in Saudi Arabia was invariably coupled with complaints about the treatment of its "neighbors," especially Iraq.[22] Placing U.S. troops in Saudi Arabia was an egregious provocation in itself, but the bases represented and facilitated the occupation of "its most powerful neighboring Arab state."[23] After 9/11, Bin Laden and his lieutenants thus threatened that the U.S. would remain a target until its military forces withdrew from the entire Persian Gulf.[24] Second, Bin Laden said the 9/11 attacks were intended to dissuade the U.S. from supporting military interventions that killed Muslims around the world. In the 1990s, this list included "Crusader wars" in Chechnya and East Timor. Actions in Israel and Iraq during this period generated the most intense opposition. After 9/11, this criticism of the U.S. focused almost exclusively on events in these two countries.[25] Third, Al Qaeda communiques emphasized its 9/11 goal of compelling the United States to stop supporting pro-Western Muslim rulers that suppress the will of their people. Al Qaeda leaders routinely denounced the House of Saud and Pervez Musharraf's Pakistan as the most "oppressive, corrupt, and tyrannical regimes" whose very existence depends on the "supervision of America."[26] A prominent Al Qaeda website equated U.S. financial and political support for Saudi Arabia and Pakistan to coloniza-tion.[27] Fourth, Al Qaeda leaders described Israel in similar terms—as a colonial outpost. Based on Al Qaeda communiques, its final objective was thus to destroy the "Zionist-Crusader alliance," which enables Israel to maintain its "occupation of Jerusalem" and "murder Muslims there."[28] In October 2001, a trove of letters allegedly written by Bin Laden was seized by Scotland Yard during an investigation of his supporters in Britain. The objectives listed in the letters were indistinguishable from those contained in his public statements: to drive out American forces from the Gulf; to deter the United States from supporting international conflicts that kill Muslims; to stop the U.S. from interfering in local politics, particularly in Pakistan and Saudi Arabia; and to end U.S. support for Israel.

September 11 not only failed to achieve any of these outcome goals; the terrorism exacerbated each of them. The 9/11 attacks were certainly counter-productive in reducing U.S. military interference in the Persian Gulf. They provided the strategic rationale for Operation Iraqi Freedom (OIF) and were the critical factor in securing the American public's support. The main selling point for regime change was not that Iraq would attack the United States or its allies, but that "using chemical, biological, or, one day, nuclear weapons provided by Iraq, the terrorists could one day kill hundreds of thousands of people in our country."[29] Postwar revelations suggest that even before September 11, 2001, President Bush and top-level officials wanted Saddam Hussein removed from power.[30] Most Americans, however, didn't support invading Iraq until August 2002, when the administration began ratcheting up its rhetoric linking Saddam to Al Qaeda. From this point until March 2003, two-thirds of the American public supported Bush's Iraq policy in the context of the wider war on terrorism.[31] Not only was September 11 a necessary condition for OIF, but the fear of emboldening the terrorists by "retreating" from Iraq became the main intellectual justification for staying there. In the absence of a major terrorist attack on the homeland, it's hard to imagine that Americans would have agreed to increase their troop presence by fifteen times in the Persian Gulf from about 11,000 before the September 11 attack to 177,000 in the course of OIF and its aftermath.[32]

The September 11 attacks also failed to destroy U.S. relations with pro-American Muslim leaders. In Pakistan and Saudi Arabia—the two bilateral relationships most objectionable to Bin Laden—9/11 led to heightened cooperation due to a combination of American pressure, inducements, and mutual fears of terrorism. Prior to the attack, U.S.–Pakistani relations were at a nadir: Pakistan had diminished strategic utility to the United States with the end of the Cold War and the Soviet occupation of Afghanistan; Islamabad posed a challenge to America's nonproliferation policy and was the principal backer of the Taliban; and General Musharraf had recently staged a successful putsch against the country's first democratically elected leader. September 11 enabled Musharraf to sell Pakistan as a critical American ally. He promptly severed diplomatic support for the Taliban and then offered intelligence, air space, and ground facilities for Operation Enduring Freedom in Afghanistan. In December 2003, the Pakistani Army and Frontier Corps began waging military operations in South Waziristan, which killed hundreds of terrorists and degraded their command and control capabilities in the region, at least to an extent. Musharraf's participation in the U.S.-led war on terrorism also included intelligence sharing and freezing assets of some terrorist entities linked to Al Qaeda.[33] In exchange, Washington stopped demanding a return to civilian rule and ignored Musharraf's periodic crackdowns on political challengers.[34] The so-called democracy sanctions were lifted, enabling Washington to provide economic and military assistance to unelected governments.[35] Sanctions against Islamabad's unauthorized 1998 decision to test nuclear weapons were likewise repealed, allowing the U.S. government to supply hundreds of millions of dollars in aid after 9/11. To boost Musharraf, president Bush and Congress also rescheduled Pakistan's $400 million debt to the United States and supported loan restructuring by the World Bank and the International Monetary Fund.[36] In these ways, 9/11 strengthened Musharraf's relations with the U.S. and bolstered his power vis-a-vis the Islamists.

The House of Saud shared several notable similarities with Pervez Musharraf before the 9/11 attack. Both rulers were allied with the Taliban, unelected, and presided over restive Muslim populations eager to curtail relations with the United States. That fifteen of the nineteen hijackers hailed from Arabia signaled to many foreign policy analysts that the "special relationship" forged between Franklin Roosevelt and Abdulaziz bin Abdulrahman Al Saud was effectively terminated. Yet 9/11 left the basic structure of the relationship unchanged. The U.S. remained dependent on their oil and the Saudis remained dependent on American consumption and security guarantees in the Gulf, particularly against Iran. Instead of driving a permanent wedge between the House of Saud and the Bush administration, the terrorist threat unified them. In late September 2001, the Kingdom severed diplomatic relations with the Taliban and began launching counterterrorism operations against local Al Qaeda operatives. A major attack in Riyadh in May 2003 demonstrated to the House of Saud its mutual vulnerability to terrorism. The Kingdom responded by detaining hundreds of suspected terrorists, killing several Al Qaeda leaders, and adopting stricter fiscal regulations to stem terrorist financing.[37] The two governments began sharing real-time intelligence on Al Qaeda movements and engaged in some joint counterterrorism operations in Saudi Arabia.[38] If anything, the 9/11 attacks seemed to provide a disincentive for removing pro-American Muslim rulers. For the Bush administration, the prospect of a Taliban-like regime assuming power in Pakistan, the Gulf monarchy countries, Algeria, and Egypt was more threatening than the risks of inflaming the proverbial Arab Street. Despite Bin Laden's stated intentions, the United States thus bolstered relations with these corrupt countries.

Bin Laden also failed to deter the U.S. from supporting military interventions that kill Muslims. The September 11 attack backfired on this objective in three ways. First, the United States responded by waging an aggressive counterterrorism campaign predominantly in the Muslim world, killing large numbers of civilians especially in Afghanistan and Iraq.[39] As part of its global campaign against Al Qaeda, the U.S. also assisted Muslim countries in capturing and killing operatives there (e.g., Pakistan, Saudi Arabia, the Philippines), with incalculable civilian losses. Second, after 9/11 the Bush administration broadened its definition of terrorism. The phrases "global war on terror" (GWOT) and "global struggle against violent extremism" (GSAVE) didn't distinguish between terrorist enemies of the United States and regional conflicts only tangentially related to the Al Qaeda threat.[40] This lack of precision was consistent with the Bush strategy of eliminating every group that intentionally attacked civilians for political gain.[41] Based on its expansive definition of terrorism, the Bush administration condoned aggressive military campaigns by other countries engaged in localized conflicts with Muslim militants. The most salient example was the Bush administration's post–9/11 orientation toward Russia. Before the attack few in the West subscribed to the Kremlin's stance equating the Chechen separatists to Al Qaeda terrorists. National Security Adviser, Condoleezza Rice, warned Russia that "not every Chechen is a terrorist" and advised the government to find a "political solution" to the "Chechens' legitimate aspirations."[42] After the attack, the Bush administration dropped its demand to find a political solution and didn't criticize President Putin when more than a thousand Chechen civilians were killed by Russian forces in the year after 9/11.[43] Similarly, the Bush administration muted its criticism of the Israel Defense Forces (IDF) in its war against Palestinian terrorists. In the four years after the 11 September attack, approximately 2,800 Palestinians were killed in Israeli counterterrorism operations, almost four times more than in the First Intifada.[44] Third, in exchange for supporting the U.S.-led war on terrorism, the Bush administration withheld criticism of American allies that had intentionally targeted Muslim civilians. After India committed itself to the war on terrorism, for example, the Bush administration decided to ignore the massacre of 2,000 Muslims in Gujarat by Hindu rioters who were reportedly supported by the Hindu nationalist government.[45] In sum, the United States not only killed thousands of Muslims in counterterrorism operations after 9/11, but the terrorist attack emboldened other governments to crack down on their Muslim opponents and dissuaded the Bush administration from containing the bloodshed.

Finally, the September 11 attacks failed to end U.S. support of Israel. Based on Bin Laden's objection to the "Zionist-Crusader alliance," one might have expected the Bush administration to downgrade relations with Israel. This seemed likely in the immediate aftermath of the attacks: In October, the president expressed his desire for Palestinian statehood; in November, Secretary of State Colin Powell for the first time mentioned the need to end the Israeli "occupation"; and in June 2002, the administration endorsed the road map for a "viable" Palestinian state.[46] These developments were superficial, though. The administration didn't object when the IDF reoccupied most of the West Bank in April 2002 and detained Yasser Arafat in his Ramallah compound. It defended the Israeli policy of targeted assassination and the construction of the security wall/fence, despite the traditional American objection to "prejudging the final status issues." This departure from American policy was codified in a Letter of Understanding between Bush and Sharon in April 2004, which formally recognized Israel's right to retain indefinitely "major population centers" (i.e., settlements) in the West Bank. Bush's pro-Israel response to 9/11 was shared by the American public. Before the attack, 41 percent of Americans expressed support for Israel and 13 percent for Palestinians. After the attack, 55 percent backed Israel, whereas support for Palestinians dropped almost by half, to 7 percent. This level of American sympathy for Israel was unmatched since the 1991 scud missile attacks. Support for a more evenhanded role in the peace process also declined.[47] To the extent Bin Laden's aim was to erode U.S.-Israel relations, 9/11 clearly failed like his other strategic demands.

Most jihadist and Islamist thinkers share my assessment. The Al Qaeda strategist, Abu Musab al-Suri, said the political outcome of 9/11 was "catastrophic."[48] Abu al-Walid al-Masri, the first non-Afghan to swear allegiance to Taliban leader Mullah Omar, said 9/11 exhibited "catastrophic leadership," as the U.S. "took an opposite turn compared to what Bin Laden had imagined."[49] Mohammed Essam Derbala, a leader of the Egyptian al-Gama'a al-Islamiya, agreed that 9/11 "produced the opposite of the desired results" since "the umma is much worse off now as a result of Al Qaeda's foolish and reckless conduct."[50] Sayyed Imam Al-Sharif, a senior Al Qaeda strategist and longtime friend of Ayman al-Zawahiri, also concluded: "Al Qaeda committed suicide on 9/11 and lost its equilibrium, skilled leaders, and influence. September 11 was a failure."[51] In his autobiography, the American jihadist, Aukai Collins, who spent time in Al Qaeda's training camps, noted that September 11 "set back the jihad movement at least two decades."[52] In Chapter 2, we'll examine the political effects of terrorism more scientifically to see whether the Strategic Model holds up any better when our sample of groups is greatly expanded.

Notes

1. Leon 1995.
2. Fielding and Penny 2006.
3. Pape 2003, 347, 350.
4. Ibid. 344; Pape 2005, 30.
5. Pape 2003, 343, 350.
6. Ibid. 354–5.
7. Crenshaw 1983.
8. Gurr 1988, 125.

9. Abrahms 2008.
10. Lake 2002; Kydd and Walter 2006; McCormick 2003, Crenshaw 2007.
11. Kalyvas 2006, 23.
12. Crenshaw 1987, 15.
13. Abrahms 2012.
14. English 2016, 1.
15. DeNardo 1985, 35.
16. Conway 2014, 204.
17. English 2016 , 115.
18. Ibid. 7.
19. Kalyvas 2006, 25.
20. Lahoud 2010, 142.
21. World Islamic Front 1998.
22. Ibid.
23. Ibid.
24. See, for example, Ibid.
25. "Bin Laden Rails Against Crusaders," November 3, 2002.
26. "Interview: Osama Bin Laden," May 1998.
27. Center for Islamic Studies and Research 2003.
28. MEMRI 2004.
29. "President's Remarks," September 12, 2002.
30. Suskind 2004.
31. Kaufmann 2004.
32. Directorate for Information Operations and Reports, 1980–2004.
33. Strategic Survey 2005.
34. Lochhead, November 11, 2001.
35. Shah 2002.
36. Hadar 2002.
37. Military Balance 2003.
38. Ibid.
39. "Iraq Body Count."
40. In July 2005, the Bush administration began employing the phrase GSAVE.
41. "President Bush Speaks," November 10, 2001.
42. McFaul 2003, 27.
43. Khan 2003.
44. B'tselem.
45. Khan 2003.
46. Strategic Survey 2005.
47. Ibid.
48. Gerges 2011, 95.
49. Ibid.
50. Ibid. 98.
51. Ibid. 121–2.
52. Collins 2002, xii.

2

Testing the Strategic Model of Terrorism

Chapter 1 suggests that the Strategic Model of Terrorism rests on a surprisingly flimsy empirical foundation. This school of thought has prominent proponents in political science. But they've offered little evidence beyond pointing to well-known cases like Palestinian terrorism against Israelis and Al Qaeda terrorism against Americans as proof that militant groups achieve their political goals by attacking civilians. Even in these cases, the civilian attacks evidently failed to redress the stated grievances. In fact, the terrorism exacerbated them. In the face of terrorism, Israelis became less supportive of territorial concessions in the Oslo Peace Process and rallied behind the security wall-fence, thereby weakening the prospects of a viable Palestinian state. Similarly, the United States responded to 9/11 by doing the political opposite of what Bin Laden demanded. The evidence against the Strategic Model becomes even stronger when we broaden our sample of militant groups.

Cherry-picking is extraordinarily easy in terrorism analysis because there have been so many incidents across space and time. That's why my research on militant groups always mixes detailed qualitative case studies, as in Chapter 1, with large-n studies, as in this chapter, to ensure my conclusions apply broadly. Equally important is to pay attention to how studies are operationalized because these decisions naturally affect the results. Methodological choices are like golf clubs. Each has its benefits and drawbacks. The putter is great for putting, less so in the rough. The sand wedge is great in the trap, but not so much for the drive. None of the analyses below purports to offer iron-clad proof that terrorism is a poor tactical choice for militant groups to attain their political demands. In the aggregate, though, they strongly support my first rule for rebels to eschew terrorism if they're serious about achieving their political goals. More specifically, I show with multiple datasets that attacking civilian targets seldom helps militant groups to attain their demands and actually discourages governments from granting them.

Testing the Strategic Model

In 2006, I published "Why Terrorism Does Not Work," the first large-n study on the political ineffectiveness of terrorism.[1] My research design was quite simple. I analyzed the political plights of twenty-eight militant groups—the complete list of foreign terrorist organizations (FTOs) as designated by the U.S. Department of State since 2001. Although the FTO designation has the word "terrorism" in it, the organizations vary considerably in their tactical choices. Some groups like the Abu Nidal Organization, Abu Sayyaf Group, Aum Shinrikyo, ETA, Hamas, Kach, and Shining Path have mainly attacked civilian targets, whereas other groups like the Revolutionary Armed Forces of Colombia, Harakat ul-Mujahidin, Hezbollah, the Islamic Movement of Uzbekistan, Mujahedin-e-Khalq, Real Irish Republican Army, and Revolution Nuclei have mainly attacked military targets. To begin assessing the political effectiveness of terrorism, I divided the organizations up in terms of whether they targeted civilians or military personnel. Although the sample size was admittedly modest, I found clear differences in the success rates of the militant groups depending on their target selection. Groups that relied on terrorism by targeting civilians systematically failed to achieve their policy demands. By contrast, groups that directed their violence against military targets were responsible for all the successful examples of government concessions. Hezbollah, for example, managed to repel the multinational peacekeepers and Israelis from southern Lebanon in 1984 and again in 2000 by attacking their troops, whereas the Shining Path failed to coerce the Peruvian government into adopting communism by killing civilians. Disaggregating the FTOs by target selection yields an important insight—terrorism has been the tactic of choice for political losers if you restrict the definition to civilian-centric violence, as most scholars do.[2]

That preliminary study was better at showing correlation than causation. Groups that rely on terrorism may have an abysmal political track record, but do attacks against civilians truly lower the likelihood of government concessions? Let's play devil's advocate for a moment. Conceivably, groups targeting civilians have a lower political success rate for reasons that have nothing to do with this tactical choice. Perhaps groups that rely on terrorism have a worse political record because they tend to be weaker than groups that target military personnel. After all, it's easier to strike soft civilian targets like a movie theater than hardened military targets like a barracks. It's also possible that civilians are more likely to be targeted when groups are up against unusually strong governments with superior resources to defend their forces. If so, then maybe the low political success rate from terrorism is a function of government capability. Another possibility is that tactically extreme groups are more likely to be politically extreme. If so, then civilian attacks are potentially correlated with political failure simply because governments are less willing to make larger political concessions, like adopting Salafism as the national religion. Such methodological challenges are tricky but not terminal for assessing the political effects of terrorism. As Ted Robert Gurr noted: "The fact that many variables affect the outcomes of violent conflict is not an insurmountable obstacle but rather a challenge for the theorist. One must specify just which circumstantial and intervening variables affect outcomes, so that they can be observed and their effects controlled for."[3] The regression analysis below enables us to tease out the independent effect of terrorism by "controlling" for alternative factors that could plausibly impact government concessions.

To assess the effect of terrorism on government concessions, I created a dataset based on the campaigns waged by every group designated before 2010 as a Foreign Terrorist Organization. This timeframe is intentional to give the militant groups sufficient time to achieve their outcome goals. My sample consists of the 125 campaigns waged by these fifty-four groups to achieve their political platforms, as specified in RAND's Terrorism Knowledge Base. Because coercion involves the application of pressure, I define a campaign as one in which an FTO wages multiple attacks that killed at least one person from the target country for the stated purpose of exacting a strategic concession.[4] The unit of analysis is therefore not the groups themselves, but their campaigns to pressure target countries into revising their policies.[5] The dependent variable is the degree to which the FTO campaigns advanced their policy demands by forcing government compliance. Coding for this variable was performed by an independent team of researchers from Cornell University applying the following criteria: a "complete success" denotes full attainment of a campaign's stated political objective; a "complete failure" indicates the absence of perceptible progress in inducing political change; and a "partial success" or a "near failure" describes middling outcomes in descending degrees of political achievement.[6] In the empirical tests that follow, I investigate the political consequences of the 125 campaigns both as a binary success–failure variable and as an ordered variable with four possible outcomes.

To test the political effectiveness of terrorism, I code each of the 125 campaigns as either terrorist or guerrilla depending on whether the target country's deaths were mainly to its civilians or military as defined by the Department of Homeland Security. Roughly half of the FTO campaigns were guerrilla in the sense that they focused their attacks on the country's military. For example, Al Qaeda in Iraq's campaign to oust the United States from Iraq is designated guerrilla because most of the Americans killed were military personnel. By contrast, the Kurdistan Workers' Party campaign against Turkey to establish an independent state there is designated terrorist because most of the

Turks killed have historically been civilians. Using a binary variable for target selection is arguably preferable to a continuous variable, as other scholars have noted.[7] Because the coercing party in international relations invariably uses a hybrid of tactics, many studies concentrate on the dominant source of pressure.[8] A binary variable is also more accurate than a continuous one because collectors of events data are less likely to miscode the dominant target selection of a campaign than to inadvertently omit particular terrorist or guerrilla attacks.[9] Indeed, there's no consensus across sources on the exact ratio of attacks against civilian versus military targets. There is a consensus, however, about whether the campaigns mainly attacked civilian or military targets. For the targeting data, I rely largely on the Global Terrorism Database at the University of Maryland, supplemented with case-specific secondary sources to ensure accuracy.[10]

The analysis below includes a battery of control variables to isolate the political effects of civilian versus military targeting or what most scholars distinguish as terrorist versus guerrilla campaigns.[11] As suggested, group capability is an important factor to take into account. In theory, groups with low capability may be able to strike civilians, but not military personnel or other government officials.[12] To ensure the results aren't driven by group capability, I control for three measures in accordance with a RAND study.[13] The first is the membership size of the group.[14] The second is the group's lifespan because older groups may develop superior organizational capacity. And the third measure is whether or not the FTOs received external support for which I employ a dichotomous variable coded 1 when they benefited from the assistance of foreign states or charities and 0 otherwise. Unlike the RAND study, I also test whether the FTO employed suicide missions because these require additional manpower and are said to boost political effectiveness.[15] For these data, I rely on RAND's Terrorism Knowledge Base, the Center for Defense Information's Terrorism Project, and Assaf Moghadam's suicide mission dataset.

The capability of the target country also merits consideration. In studies of economic sanctions, countries with greater capability are less susceptible to coercion.[16] The same causal logic potentially applies against militant groups. I operationalize four variables to test the capability of the target countries. First, I use the World Bank's World Development Indicators and the Center for International Comparisons' Penn World Table to test the logged per capita gross domestic product for the year in which the target country suffered the greatest number of campaign-related deaths. Second, I use data from the World Penn Table to test the logged populations of the target countries. Third, I test the fighting capability of the target countries with data from the Correlates of War's Composite Index of National Capability, an aggregate indicator of military power (e.g., army size) and economic power (e.g., energy consumption). Fourth, I use Polity IV to control for the regime type of the target country because democracies are often seen as worse at counterterrorism than authoritarian countries because of their stronger inhibitions against crushing dissent.[17]

Finally, I control for the nature of the political demand because the greater the anticipated cost to the government of granting it, the lower the likelihood of compliance.[18] Like parents, governments are more likely to grant smaller demands than larger ones. Some scholars believe that the size of the demand is the key variable affecting the likelihood of government concessions.[19] There's a general consensus within international relations on how governments rank political objectives. A campaign defined as having a "maximalist" objective is fought to induce the target government to cede power or alter its ideology, whereas a campaign defined as having a "limited" objective is fought to achieve ends that wouldn't directly imperil the government or its citizens' fundamental way of life.[20] Examples of the former include the Mujahedin-e Khalq's campaign against Iran to end clerical rule, and the Popular Front for the Liberation of Palestine's campaign against Israel to establish Marxism in historic Palestine. Examples of the latter include ETA's campaign against Spain to permit the creation of a relatively small Basque state within that country, and the Continuity Irish Republican Army's campaign against the United Kingdom to allow Northern Ireland to secede. To test the effect of the campaign's political objective, I employ a dichotomous variable coded 1 if the objective is maximalist and 0 if it's limited.

Are Terrorist Campaigns Politically Effective?

Militant groups sometimes triumph. In the sample of 125 FTO campaigns, 38, or about 30 percent, successfully coerced the target country into at least partially complying with the policy demand. Figure 2.1 adds important insight, demonstrating a clear difference in the political success rates of guerrilla and terrorist campaigns. The number of FTO campaigns that targeted a country's military personnel was about the same as the number that targeted civilians: 60 and 64, respectively, or 48 and 51 percent. But the guerrilla campaigns targeting a state's military account for 36 of the successful cases of coercion. The lone case in which a terrorist campaign even partially achieved its policy demand was the highly publicized Spanish decision to withdraw from Iraq in response to the March 11, 2004, Madrid train station bombings.[21] In this special case, an Al Qaeda-linked group called the Moroccan Islamic Combatant Group attacked Spanish commuters days before the prime ministerial election, which may have helped to elect the antiwar candidate, José Luis Rodríguez-Zapatero, who then fulfilled his pledge to bring the Spanish troops home. The data suggest, however, that the so-called 11-M campaign is an outlier. Target selection matters greatly across

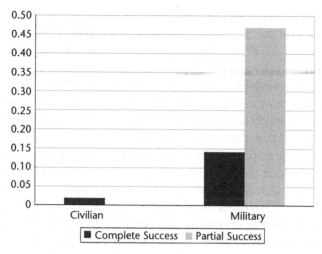

Figure 2.1. Proportion of Successes by Target

model specifications. When FTOs rely on a guerrilla campaign, it often succeeds in achieving their stated political goal. But when these same groups rely on a terrorist campaign, it nearly always fails regardless of their capability, that of the target country, or even the nature of their political demands. Terrorism is hence an ineffective tactic for militant groups to achieve their demands.

Table A.1 in the Appendix displays the results of four multivariate regression models. Column 1 presents results of a standard logit model using a binary variable for success or failure of the FTO campaigns: a complete success or partial success is coded as 1, and a complete failure or near failure is coded as 0. In cases where a group wages multiple campaigns, these may not be fully independent of each other, so I also calculate robust standard errors with observations clustered by FTO. Holding other factors fixed, the target selection of an FTO campaign is highly significant and substantively more important than any other explanation. The nature of the political objective has a fairly large effect in the expected direction; that is, campaigns fought to achieve maximalist goals are predictably less likely to succeed. The size of FTO membership is also significant but substantively meaningless and the coefficient is in the wrong direction, a function of Al Qaeda's status as a large, ineffective outlier.[22] When campaigns from this FTO are dropped, group size has no statistically significant effect. No other variables achieve statistical significance. Even with this admittedly blunt coding of the dependent variable, the evidence strongly suggests that campaigns directed against civilians are counterproductive for coercing governments into making concessions.

In Column 2 of Table A.1, I add nuance to the dependent variable by employing an ordered logit model with four campaign outcomes: complete success, partial success, near failure, and complete failure. In this improved version, I include the full set of independent variables from the benchmark model and again calculate robust standard errors with observations clustered by FTO. The results remain the same: targeting civilians has a large, significant negative effect on campaign outcomes. The odds of this result occurring by chance are less than one in a thousand. As expected, campaigns fought to achieve maximalist objectives are also significantly less likely to be successful, but the impact of target selection is even greater.

To interpret the impact of the significant variables, I calculate in Table A.2 their marginal effects coefficients, holding all other variables fixed. Row 1 shows that an FTO campaign is 77 percent more likely to completely fail when directed against a country's civilians as opposed to its military. Conversely, when a campaign is directed against a country's military rather than its civilians, the odds of achieving at least partial success increase by 55 percent. Row 2 presents the marginal effects coefficients for the nature of the campaign's political objective, as a means of comparison. When a discrete change is made from limited to maximalist objectives, the odds of complete failure increase by 40 percent, and the odds of obtaining at least partial success decrease by just 22 percent. Furthermore, when the target is civilian, not military, the odds of achieving complete success decrease by 10 percent, though this figure declines by only 2 percent when the campaign objective is maximalist as opposed to limited. When it comes to the political success rates of the campaigns, target selection is an even more consequential factor.

Figure 2.2 illustrates visually the change in outcome probabilities for the two statistically significant independent variables: the campaign's target selection and political objective. A change in targeting from military to civilian generates high positive odds for complete failure, and the likelihood of partial success or even near failure falls sharply. In other words, whereas guerrilla campaigns are a productive way for groups to compel at least partial compliance, terrorist campaigns are an almost surefire way to obtain no concessions at all.[23] The campaign's political objective follows a similar pattern: adopting maximalist goals increases the odds of complete failure and decreases the chance for any level of political success. Yet the magnitude of change is clearly weaker than for the campaign's target selection. In sum, across model specifications the tendency for governments to comply depends crucially on the campaign's targeting. The evidence strongly suggests that rebel leaders would be wise to refrain from engaging in terrorist campaigns against civilians.

These findings were subjected to numerous robustness checks. First, because Al Qaeda and its affiliates are sometimes regarded as *sui generis*, I dropped their campaigns from the dataset and recalculated the ordered logit model shown in Column 2 of Table A.1.[24] This decision simultaneously addresses the spurious significance of group membership size. Column 3 presents the results: the

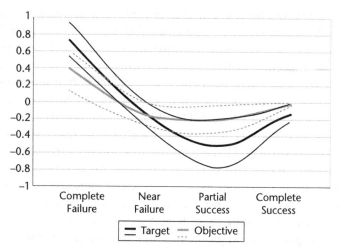

Figure 2.2. Change in Probability of Campaign Outcomes

campaign's target selection remains both statistically and substantively the most significant variable. Second, I re-ran the models, dropping the campaigns in Afghanistan and Iraq because of the fluidity of these theaters. The results, presented in Column 4, remain unchanged.[25] Third, I examined whether the poor success rate of terrorist campaigns is driven by the fact that they're more likely than guerrilla campaigns to take place inside the target country. The effect of distance is generally neglected in political science, but to wage distant campaigns governments incur additional costs of supply and transport.[26] The findings stand after controlling for whether the FTO campaigns were waged at home or abroad. Fourth, to confirm that terrorist campaigns are politically ineffective even when they employ suicide attacks, I reran the models interacting the FTO campaigns that used suicide missions and targeted civilians. The interaction term is insignificant, and targeting civilians remains both statistically and substantively the most important variable. Fifth, I reran the models lagging the independent variables pertaining to the target country's capability in order to allay concerns of potential endogeneity; testing them at the start year of the campaign rather than at its peak year of violence does not attenuate the findings. Sixth, I examined whether guerrilla campaigns tend to kill more people than do terrorist campaigns because that tendency would arguably account for their variable coercion rates; in fact, the number of civilians killed in terrorist campaigns generally exceeds the number of military personnel killed in guerrilla campaigns.

Finally, I paid extra attention to the possibility of a selection effect in which terrorist campaigns fail because only weaker groups wage them. To address this potential concern, I disaggregated the data into campaigns waged by small and large FTOs, using the median of 1,000 members as the breakpoint.[27] Their overall success rates are similar: small groups achieved partial or complete success in 30 percent of their campaigns compared to 29 percent for large groups. More important, both sized groups struggled to coerce in their terrorist campaigns. For small FTOs, their guerrilla campaigns often succeeded, but all 51 of their terrorist campaigns failed, except for the Spanish outlier. The results are equally stark for large FTOs: all 11 of their terrorist campaigns failed. I also applied the fully-specified ordered logit model discretely to the campaigns waged by small and large FTOs, using a dummy variable to control for group membership size.[28] In all of these analyses, target selection remains the key determinant of campaign success, both statistically and substantively. As a further check, however, I also examined the political success rates of the FTOs that engaged in both guerrilla and terrorist campaigns. Among these multi-tactical groups, an appreciably lower success rate in their campaigns against civilians would suggest that target selection matters independently of group capability. Indeed, these multi-tactical groups achieved at least partial success in 31 of their 46 guerrilla campaigns, while failing in their ten terrorist campaigns. In other words, when an FTO wages a guerrilla campaign, it's frequently successful in coercing at least partial compliance. When that same group wages a terrorist campaign, however, it fails nearly systematically.

For several reasons, this analysis may even overestimate the ability of terrorist campaigns to coerce strategic concessions. First, FTOs are endowed with far greater capabilities than most other groups that use terrorism. Because my sample is restricted to the campaigns of FTOs, it's a "hard test" for my thesis about the political ineffectiveness of civilian attacks. Second, the literature on coercion emphasizes that it's more likely to succeed when a campaign is fought to obtain a clearly articulated political demand and all 125 campaigns meet this criterion.[29] This decision rule excluded from analysis a large number of terrorist campaigns that have been fought to achieve unidentifiable political ends, which are essentially unwinnable.[30] Third, case studies illustrate that terrorism can exacerbate the perpetrators' grievances as we saw in Chapter 2. Although backfiring is worse for the coercing party than the null result, none of the models assesses additional penalties for politically counterproductive terrorist campaigns.

Testing the Effects of Terrorism in Hostage Situations

The foregoing analysis provides strong evidence that militant groups should refrain from terrorism to achieve their political goals. The robust empirical relationship between civilian-centric campaigns of violence and political failure appears to be causal rather than an artifact of other factors like the capability of the perpetrators or the nature of their demands. But all methodological approaches have limitations. The first pertains to coding the independent variable of target selection. Arguably, disaggregating the FTO campaigns in terms of whether they mainly struck military or civilian targets obscures their discrete tactical effects. One might reasonably wonder whether civilian attacks in military-centric guerrilla campaigns helped to pressure the concessions or whether military attacks in civilian-centric terrorist campaigns somehow adversely affected the political outcome. The broader problem is that campaigns tend to employ a hybrid of tactics, making it difficult to tease out the political impact of attacking one target over another. To exact political concessions from Israel, for instance, Fatah, Hamas, and Hezbollah have attacked both the population and military, while concurrently underwriting anti-Zionist civil resistance initiatives. In fact, many militant groups have a political wing that participates democratically. Japan's Aum Shinrikyo, for example, selected candidates to contest local elections while dispersing Sarin gas in the Tokyo subway. Discerning the political consequences of such tactics is admittedly difficult when employed in tandem.[31]

The second methodological limitation pertains to coding the dependent variable. Historically, datasets have neglected to code the political outcomes of asymmetric campaigns, the standard unit for assessing the tactical value. To compensate for this lacuna, scholars testing the effectiveness of terrorism have themselves coded the outcomes of the campaigns, inviting allegations of confirmation bias.[32] Scoring the extent to which militant groups accomplish their political ends is objectively difficult. Terrorists are notorious for issuing protean, ambiguous political demands or sometimes none at all.[33] Disagreement over terrorism's effectiveness can therefore hinge on mini-empirical disputes over whether the perpetrators accomplished their desired strategic goals. In recent years, such dissension has plagued the coding of whether Al Qaeda, its affiliates, the Irish Republican Army, its splinter groups, Palestinian terrorists, the Tamil Tigers, and its organizational rivals have realized their political goals, muddying assessments of terrorism's overall tactical value.[34] Adding to the confusion over coding the dependent variable is that militants may possess unusually long time horizons. Lashkar-e-Taiba, for example, has so far failed in its stated mission of spreading Islamic rule throughout India. But the group remains operational and may one-day wrest control over Indian Kashmir. Although terrorist campaigns may persist for decades without any perceptible political return, this lengthy timeframe is perhaps acceptable for those committed to the causes of the armed struggle. Methodologically, scholars have dealt in an ad hoc way with these ongoing campaigns by excluding them from the analysis, which artificially drives up the coercion rate, or by including them in the analysis, driving down the coercion rate.[35] Scoring political progress is also problematic if terrorists express unrealistic demands in order to obtain even a fraction of them. Bin Laden's 1998 Fatwa, for example, called on the United States to withdraw from the Middle East. This demand remains unmet, of course, but Al Qaeda may deem the smaller concession of withdrawing from Saudi Arabia five years later as an important victory in itself. Similarly, the Revolutionary Armed Forces of Colombia may have been content to command southeastern parts of the country despite failing to enact Bolivarianism as the official national ideology. Indeed, militant groups tend to fall short of accomplishing their strategic demands, but may achieve some measure of progress in the form of partial government accommodation. An ordinal dependent variable can help to capture such middling levels of bargaining success. Yet weighting the political outcomes inevitably introduces an element of subjectivity.

These methodological issues can be sidestepped with a different empirical strategy. International Terrorism: Attributes of Terrorist Events (ITERATE) contains fine-grained data on over a thousand international hostage situations between 1968 and 2005 in which militant groups adopted a variety of tactical approaches to pressure governments into accommodating their demands. ITERATE furnishes tactical data in terms of whether the hostage-takers killed or wounded civilians, government officials, both kinds of captives, or nobody at all, as well as the outcome of the negotiations. This information provides additional leverage for analyzing the effects of various rebel tactics in a relatively controlled bargaining setting.

The demands issued in hostage settings are redemptive rather than strategic. Whereas the strategic demands of terrorist groups reflect the political aspirations of their leaders, redemptive demands are designed to sustain the organization by eliciting financial and other material concessions.[36] Redemptive demands include financial ransom, prisoner release, or safe passage out of a location or to a specific destination. All of these demands are clear and explicit, the bargaining outcomes hail directly from ITERATE, and none of the confrontations is ongoing, facilitating objective assessment of whether the government acquiesced. I code each confrontation as either a success or a failure depending on whether the perpetrators received any of their demands.

Because consensus remains lacking over the definition of terrorism, this variable is coded in multiple ways. Model 1 employs a variable for terrorism based on a loose interpretation of the definition. In this model, the tactic is coded as present for incidents in which the hostage-takers killed or wounded civilians, government officials, or both sets of captives.[37] This dummy variable simultaneously captures an alternative tactical approach—the decision to not physically harm anyone at all. As Martha Crenshaw and others note, the decision to use terrorism is never the only available option because refraining from physical harm is always an alternative.[38] Model 2 codes the variable of terrorism based on a narrower, more common definition by incorporating only incidents in which the perpetrators killed or wounded civilians in particular. Model 3 includes a separate tactical variable for incidents that killed or wounded only security forces, military personnel, or other government officials, a distinction from terrorism that's becoming the norm.[39] This empirical analysis thus departs from the previous research design by adding specificity to the tactical variable of terrorism and comparing its efficacy with available alternatives in which neither the means nor ends are disputable.

As before, all models include a raft of control variables to help isolate the independent tactical effects, especially the value of harming civilians for pressuring government concessions. I again factor out the capability of the government and perpetrators as well as the type of demand issued. I also control for the number of demands issued in the stand-off because economic theories on bargaining imply that additional demands boost the chances of at least one being granted.[40] Despite the richness of these data, several of the variables lack numerical values. To account for these data, Amelia II is used to multiply impute them. This statistical program applies a bootstrapping algorithm that imputes m values for each missing cell in the data matrix. By doing so, data from all 1,075 reported hostage incidents are preserved. In cases where the same group engages in multiple confrontations over time, these may not be fully independent, so robust standard errors are included, thereby correcting for heteroskedasticity and autocorrelation.

Is Terrorism Effective in Hostage Situations?

Table A.3 in the Appendix presents the results. The evidence again strongly indicates that militant groups benefit from tactical restraint. Model 1 estimates that the decision to kill or wound civilians, government officials, or both kinds of captives significantly lowers the likelihood of bargaining success ($p < 0.05$). By ensuring the physical safety of others, hostage-takers are more likely to induce government accommodation. Model 2 adds analytic clarity by employing a narrower definition of terrorism. In this model, the tactical variable of terrorism is restricted to cases in which only civilian captives are killed or wounded. Harming civilians in particular has an independent, negative impact on the likelihood of government accommodation ($p < 0.05$). By contrast, Model 3 illustrates that harming military personnel and other government officials doesn't have a significant effect in impeding concessions. There is thus no empirical evidence that terrorism is a winning tactical decision for aggrieved groups to obtain their demands. On the contrary, escalating against the captives is consistently the worst tactical option especially when civilians are the target.

To estimate the substantive impact of terrorism on government intransigence, I calculate the marginal effects of employing this tactic holding all other variables at their mean. Table A.4 displays the predicted probabilities of negotiation success and the 95 percent confidence intervals for each point estimate. Compared to refraining from physically harming any of the hostages, engaging in terrorist violence lowers the chances of bargaining success by 6 to 7 percentage points depending on whether civilians are the only ones killed or wounded in the hostage incident. The marginal effect of capturing additional hostages is also calculated to put this level of impact into greater perspective. This is a useful comparison because many scholars theorize that the success of militant groups depends at least as much on their capabilities as on their tactical decisions.[41] Yet refraining from killing or wounding civilians has a relatively large substantive impact, the equivalent of the perpetrators seizing 99 hostages per incident, roughly six times the mean and twice the standard deviation. For achieving the most testable type of terrorist group demands, the decision to escalate is folly.

Numerous robustness tests were performed to ensure that the tactical results aren't a quirk of the econometric specifications. Skeptics might wonder, for example, whether the findings are due to reverse causation in which the perpetrators escalate to terrorism after the government refused to accommodate their demands. To address this potential objection, Model 4 in Table A.3 includes an important dummy variable called "Negotiation Harm" that controls for whether the terrorist violence was perpetrated during the bargaining process. This variable isn't statistically significant; killing or wounding civilians is evidently exogenous to bargaining failure, consistently lowering the odds of government compliance regardless of when the tactical escalation occurs in the course of the standoff. A final concern to address is whether just a handful of groups are driving the statistical results. To attend to this potential issue, all of the models have been rerun using multilevel logistic regression taking into account individual group factors. There's virtually no change in the coefficients for the tactical variables. Killing captives particularly civilian ones decreases the probability of government concessions regardless of the specific hostage-takers.

In sum, my research consistently finds across multiple data sources that killing civilians is an ineffective, even counterproductive tactical decision for groups to attain their demands. This observation holds true whether the unit of analysis is the dominant target selection of the group, the campaigns it wages, or is even down to the attack-level in hostage settings. Each approach has benefits and drawbacks. In the aggregate, these large-n analyses provide strong additional evidence that rebels should think twice before using terrorism to elicit government concessions.

My original finding is strengthened by the fact that other researchers have run their own tests based on different research designs and reached complementary results. The political scientists Page Fortna and Jessica Stanton find support for my thesis that groups lower the odds of government concessions by attacking the population with terrorism.[42] In another study, Erica Chenoweth and Maria Stephan find that protest groups suffer at the bargaining table when they engage in violence, including against civilians.[43] Anna Getmansky and Tolga Sinmaz-demir analyze the relationship between Palestinian violence and Israeli outposts in the West Bank between 2000 and 2005. Consistent with my experience in the West Bank during the Second Intifada, they find that Palestinian attacks lead to a proliferation in the number of Israeli outposts contra the political objectives of the perpetrators.[44] Also in accordance with my finding is the evidence terrorism tends to boost popular support for right-wing leaders opposed to appeasing the perpetrators. In a couple of statistical studies, for example, the economists Claude Berrebi and Esteban Klor demonstrate that Palestinian terrorist attacks increase Israeli support for the Likud and other right-wing parties opposed to territorial concessions.[45] The most lethal Palestinian attacks are the most likely to induce this rightward electoral shift.[46] This trend is hardly unique to Israel. Christophe Chowanietz analyzes public opinion within France, Germany, Spain, the United Kingdom, and the United States from 1990 to 2006. In each country, attacking civilians has shifted the electorate to the political right, strengthening the hand of hardliners most opposed to appeasement.[47] Similar observations have been noted in at least nine other countries after Al Qaeda and its affiliates harmed their citizens.[48] RAND summarizes some of this research: "Terrorism fatalities, with few exceptions, increase support for the bloc of parties associated with a more-intransigent position toward terrorism and territorial concessions . . . Some scholars may interpret this as further evidence that terrorist attacks against civilians do not help terrorist organizations achieve their stated goals."[49] Another study in the *American Political Science Review* finds that even a heightened threat of terrorism to a certain locality may increase popular support for right-wing leaders.[50] Finally, several statistical studies have found that attacks on civilians reduce the odds of concessions when perpetrated by national militaries.[51] Together, this body of research indicates that civilian attacks have an independent, negative effect on the likelihood of government concessions. Rebels interested in gaining government concessions are wise to refrain from terrorism.

Notes

1. Abrahms 2006.
2. Ibid.
3. Gurr 2015, 150.
4. Global Terrorism Database.

5. The number of campaigns exceeds the number of FTOs because many of these groups have waged multiple campaigns. Ansar al-Islam, for example, has waged two campaigns: one against Iraq to establish Sharia law, and one against the United States to expel it from Iraq.

6. To illustrate, an example of a complete success is Hezbollah's expulsion of American peacekeepers from Lebanon in 1984; a complete failure is Aum Shinrikyo's lack of progress in establishing its syncretic belief system in Japan; a partial success is the Revolutionary Armed Forces of Columbia's accomplishment of controlling substantial territories throughout that country; and a near failure is ETA's weak record in establishing an independent Basque state in Spain. For uniformity in the coding process, I supplied more specific instructions for evaluating campaigns to eject occupying countries: a complete success signifies the total withdrawal of the target country; a complete failure describes a campaign in which the target country did not reduce its troops; a partial success refers to a campaign in which the target country has either reduced its troops or officially expressed its intentions to do so; and a near failure describes a withdrawal from less than 10 percent of the contested territory, with no official declaration to terminate the occupation.

7. Stanton 2016, 67; see also Kalyvas 2006.

8. George and Simons 1994; Art and Cronin 2003.

9. Drakos and Gofas 2006.

10. For one campaign, the Khmer Rouge's to replace the Lon Nol government, the target selection is undetermined: GTD data on the group begins too late, in 1978, after the group had assumed power. And there's no consensus in the historical literature on whether the group attacked mostly civilians or military personnel en route to power.

11. See Allemann 1980; Kossoy 1976, Richardson 2007.

12. See Findley and Young 2012; Crenshaw 1981.

13. Jones and Libicki 2008.

14. Group memberships ranged in size from 40 (Japanese Red Army) to 50,000 (Al Qaeda).

15. Pape 2003.

16. Hart 2000.

17. Abrahms 2007.

18. George and Simons 1994.

19. Pape 2003.

20. Abrahms 2006; George and Simons 1994; Marinov 2005.

21. Adding the numbers of successful guerrilla campaigns to successful terrorist campaigns does not equal 38 due to the missing targeting data for the Khmer Rouge's campaign.

22. Al Qaeda-in-Iraq, a different FTO, has been comparatively successful.

23. Complete success is hardest to achieve, so the change in probabilities generated from changes to these independent variables do not substantially differ from zero.

24. Crenshaw 2008.

25. This robustness check concurrently preempts a potential objection with my research design—that guerrilla campaigns are more likely to succeed because sometimes

multiple groups wage them in pursuit of a common political goal. In my sample of substate campaigns, Afghanistan and Iraq are the only theaters that conform to this scenario.

26. Mearsheimer 2001.
27. As already demonstrated, none of the measures for group capability is significant. I focus here on membership size because this is the most common measure of group strength. In fact, the U.S government often relies on the membership size of the group as the sole indicator of its capability (U.S. Department of State, 1999; see also Center for Defense Information, 2009). When an FTO had 1,000 members, I coded the group as small because the next smallest group had 3,000 members, suggesting a more natural breakpoint after, rather than before, the 1,000-member median.
28. The model is unable to converge for the large FTOs due to excessive campaign clustering. When FTOs with 1,000 members are coded as large, the model converges and the coefficient on civilian targeting remains negative and significant ($p < 0.001$). The number of campaigns waged by small and large FTOs totals 121, not 125, due to missing data for the membership sizes of Ansar al-Sunnah and Hizbul Shabaab.
29. George and Simons 1994.
30. Schelling 1991.
31. Weinberg, Pedahzur, and Perliger 2003.
32. Rose, Murphy, and Abrahms 2007; Chenoweth et al. 2009; Moghadam 2006; Krause 2013.
33. Roy, Burdick, and Kriesberg 2010.
34. Pape 2003; Abrahms 2006.
35. Jones and Libicki 2008.
36. Gambill 1998.
37. This is a minority interpretation of the definition, but not without adherents. See, for example, Pape 2003; Kydd and Walter 2006.
38. Crenshaw 1992.
39. Abrahms and Potter 2015.
40. Sandler and Scott 1987.
41. DeNardo 1985; Lake 2002.
42. Fortna 2015; Stanton 2017.
43. Stephan and Chenoweth 2008.
44. Getmansky and Sinmazdemir 2018.
45. Berrebi and Klor 2006; Berrebi and Klor 2008.
46. Gould and Klor 2010.
47. Chowanietz 2009.
48. See, for example, Mueller 2006.
49. Berrebi 2009, 189–90.
50. Getmansky and Zeitoff 2014.
51. Downes 2007.

3

Terrorism Success Stories Revisited

Thus far, I've presented some multi-method evidence in support of my first rule for rebel leaders of avoiding terrorism to achieve their political ends. Last chapter, we saw the political ineffectiveness of terrorism on large samples of militant groups from multiple, independently collected global datasets going back several decades. Large-n studies are essential for making this assessment because a few cases are insufficient for establishing trends. And regression analysis enables us to carefully tease out the independent effects of terrorism on government intransigence across time and space. Together, these tests strongly suggest that terrorism is not just a strategy for losers, but a losing strategy. That is, civilian attacks aren't just correlated with political failure; they help to cause it.

Still, it's useful to scrutinize the most commonly cited cases of terrorist successes to truly appreciate their rarity. In Chapter 1, we saw how proponents of the Strategic Model count Palestinian terrorism during Oslo and Al Qaeda terrorism on 9/11 as proof that civilian attacks pay, when both terrorist campaigns backfired on the perpetrators politically. Other asymmetric campaigns are also commonly invoked in which the militant groups did in fact manage to achieve their stated political goals. If even these apparent terrorism success stories fail to illustrate the political value of civilian targeting then they would strengthen the evidentiary case for rebel leaders to oppose this tactic. This chapter shows that the most frequently referenced historical examples of terrorist victories are indeed problematic. Arguably, these political successes weren't due to terrorism at all.

Hezbollah, Irgun, and African National Congress

If you go to the terrorism books section in the library stacks, you'll invariably encounter a massive literature on the definition of terrorism. If there's one thing every terrorism scholar seems to agree upon it's that no consensus exists over what terrorism means. This lack of definitional consensus may seem like a perfunctory, annoying semantic point. And it can be. What's important though is how this confusion over the definition of terrorism clouds perceptions of its tactical value. I've introduced the distinction between what I call terrorism splitters versus terrorism lumpers.[1] Splitters define terrorism narrowly, as the use of violence against civilian targets in particular. Lumpers, by contrast, employ an expansive definition of terrorism, brooking no distinction between this tactic and guerrilla attacks against military and other government targets. Most terrorism scholars are splitters. As Paul Wilkinson observed decades ago, "It is an elementary but common mistake... to equate terrorism with guerrilla war."[2] His seminal history of terrorism points out that "guerrilla warfare is often confused with terrorism."[3] This distinction is relevant because when people highlight cases of terrorism "working" they're usually guilty of over-aggregating or lumping. In Pape's sample of successful "terrorist" campaigns, for example, most of them were directed military targets rather than civilian ones.[4] As Jeff Goodwin points out, "It is a mistake, therefore, to refer to these campaigns as terrorist in nature, as Pape does."[5] Below, I show that terrorism wasn't responsible for the most celebrated militant groups victories because they relied on alternative tactical pressure, at least according to splitters like me. To be clear, what matters isn't how you or I wish to define terrorism. My point is that the most salient examples of "terrorism" winning didn't pressure governments into complying by attacking their civilians.

Researchers like to cite Hezbollah's success in coercing Western forces from Lebanon as prima facie evidence that terrorism works. Two truck bombs in Beirut killed 241 Americans and fifty-eight Frenchmen on October 23, 1983, pressuring the multinational peacekeeping force to withdraw from the country. As the terrorism experts, Bruce Hoffman and Matthew Levitt, note in separate studies, this withdrawal from Lebanon is widely cited as the strongest example that "terrorism works."[6] Malcom Nance, for instance, says this withdrawal proved to him that "terrorism is effective."[7] The political scientists, Andrew Kydd and Barbara Walter, likewise hail this salient case as evidence that "terrorism often works."[8] Nobody disputes that the pain inflicted by Hezbollah persuaded the Americans and French to get out of Lebanon. But were the truck bombings really terrorist incidents? Not if terrorism is to mean violence against civilians. The target of the attack was a barracks. Americans lost 220 Marines, eighteen sailors, and three soldiers in the deadliest single-day death toll for the U.S. Armed Forces since the Vietnam War.[9] The French lost fifty-five paratroopers from the 1st Parachute Chasseur Regiment and three paratroopers of the 9th Parachute Chasseur Regiment in the worst national military loss since the end of the Algerian War.[10]

Hezbollah successfully forced not only the multinational peacekeeping force out of southern Lebanon in the early 1980s, but also the Israel Defense Forces in 2000. Again, though, the target of the pressure was military—not civilian. Hezbollah specialist Daniel Sobelman notes that the attacks targeted the IDF Security Zone, with the understanding that "under no circumstances will civilians be the target of attack."[11] In his book on Hezbollah, the anthropologist Augustus Richard Norton details how the "resistance operations were mainly targeted against Israeli soldiers . . . not against civilians."[12] Of course, Hezbollah doesn't always follow these rules of engagement. In July 2006, the Party of God fired thousands of Katyusha artillery rockets into northern Israel, killing forty-three civilians. And in July 1994, the group blew up the Buenos Aires Jewish Community Center in July 1994, killing eighty-five.[13] The point isn't that Hezbollah always forswears terrorism, but that the so-called "A-Team of Terrorists" successfully coerced the American, French, and Israelis from southern Lebanon by attacking their military forces—hardly strong examples of "terrorism" working as most people understand the term.[14]

The British withdrawal from Mandatory Palestine in May 1948 is another commonly cited example of terrorism's vaunted effectiveness. The Irgun attacked the British, leading to their retreat and ultimately the creation of the State of Israel. In his book on terrorist outcomes, the historian Richard English flags this violent episode as among "the past's best candidates for a terrorist campaign which centrally worked."[15] The economists, Walter Enders and Todd Sandler, also claim that this Irgun victory "demonstrated the effectiveness of urban terrorism worldwide."[16] Max Boot says Irgun violence against the British amounts to "one of the most successful terrorist campaigns ever."[17] Not only did the asymmetric violence coerce the British to abandon Palestine and redound to Israeli statehood, but Irgun leader, Menachem Begin, would later go on to become prime minister. As Walter Laqueur points out, though, "Terrorism was not the decisive factor" in the British withdrawal.[18] Notably, the Irgun targeted British troops—not civilians. A detailed book-length history of this case concludes, "Unlike many terrorist groups today, the Irgun's strategy was not deliberately to target or wantonly harm civilians."[19] There were select cases of indiscriminate violence against the British to be sure, but Begin opposed them.

On July 22, 1946, Begin ordered a hit on the King David Hotel in the most notorious terrorist incident of the conflict. Disguised as hotel waiters, Irgun operatives planted a bomb in the basement of the hotel, killing ninety-one people, mostly civilians: forty-one Arabs, twenty-eight Britons, seventeen Jews, as well as two Armenians, a Russian, an Egyptian, and a Greek national.[20] The dead included hotel employees and typists, canteen workers, and five members of the public who happened to be in the hotel or on the street when the bomb detonated.[21] But this hotel wasn't a normal hotel. It served as the headquarters for the British Armed Forces in Palestine. By all accounts the intent wasn't to harm civilians. Hoffman notes, "The attack's target, therefore, was neither the hotel itself nor the persons working or staying in it, but the government and military offices located there."[22] As Begin recounts in his memoire, "There were many civilians in the hotel whom we wanted, at all costs, to avoid injuring."[23] Indeed, the Irgun had repeatedly warned civilians to evacuate before the attack. The Irgun placed multiple telephone calls to the hotel dispatch, the *Palestine Post* newspaper, and the French Consulate warning of the bomb according to Palestinian and U.S. officers at the time. Tragically, the message wasn't followed.[24] The attacks on civilians were not only unintentional, but counterproductive. Far from illustrating the strategic benefits of terrorism, the "collateral damage" nearly destroyed the Irgun and its Zionist aspirations. The British responded to the hotel attack with Operation Shark—the largest search-and-seizure exercise against the group. The terrorism specialist, Audrey Cronin, describes the fallout from the hotel bombing, "Public reactions to the bombing were universally critical, including among the Jewish population in Palestine, and the backlash nearly led to the destruction of the Irgun."[25] The Haganah, the principal Jewish group in Palestine, blasted the civilian casualties because they only "facilitate the government's war on the Zionist enterprise and hinder our crucial struggle."[26] The withdrawal from Mandatory Palestine is thus hardly a strong example of terrorism paying because the Irgun targeted military personnel, tried to spare British civilians, and the accidental carnage only seems to have put Zionist goals at risk.

Perhaps even more than Hezbollah and the Irgun, the African National Congress is often championed as a terrorism success story for ending South African apartheid. The legal scholar, Alan Dershowitz, says the ANC's triumph is perhaps the strongest evidence that "terrorism works."[27] The sociologist, Clifton Bryant, says the ascent of its leader, Nelson Mandela, shows that "yesterday's terrorists often become today's peacemakers and tomorrow's prime ministers."[28] As Louise Richardson points out, though, "Mandela led a campaign of sabotage, not terrorism."[29] Goodwin adds, "The ANC simply did not carry out much terrorism."[30] This is certainly true if we restrict the definition of terrorism to ANC attacks against civilians, which were exceedingly rare, contrary to the leadership's targeting strategy, and politically counterproductive.

The ANC's plight supposedly demonstrates the political utility of terrorism, but Mandela never wavered about the strategic benefits of sparing civilians. In courtroom testimony at his 1964 trial, he explained that operatives were "meant to attack only political, military, and economic targets, with minimal loss of life, in an effort to change government policy."[31] The judge agreed that Mandela promoted such targeting selectivity.[32] Even in private, the leadership instructed operatives against harming civilians. The police recovered an ANC

manual in July 1963 known as Document 73 with detailed targeting guidelines. There was to be "a massive onslaught on pre-selected targets," but none
civilian.[33] Specifically, the manual instructed operatives to attack only
"(I) strategic roads, railways and other communications; (II) power stations;
(III) police stations, camps and military forces; (IV) irredeemable Government
stooges."[34] When apartheid began to crumble in the late 1980s, the *New York
Times* confirmed that "it has remained the group's policy not to deliberately
attack civilian targets."[35] For reasons that will become clear later in the book,
operatives didn't always follow the leadership's targeting guidelines. In the
1980s, the bombing of the Wimpy bar outside Johannesburg killed five
civilians and the bombing of the Magoo bar on Durban beach killed three
civilians.[36] Rather than helping their political cause, though, ANC leaders saw
these unwanted incidents as counterproductive. Mandela writes in his autobiography, "Terrorism inevitably reflected poorly on those who used it, undermining any public support it might otherwise garner."[37] Other ANC leaders
agreed that civilian attacks "had a devastating political effect" that "consoli-
dated white opinion" and "strengthened the hand of the white government,"
as "the enemy took what we were doing and tried to use it against us."[38] If
anything, the ANC triumphed despite its smattering of terrorist attacks rather
than because of them.[39]

In sum, the main historical examples offered up by researchers that terrorism
pays—the withdrawal of Western and Israeli forces from Lebanon, the departure of Britain from Palestine, and the ending of apartheid in South Africa—are
arguably not examples of terrorist campaigns at all. All three groups coerced
governments to concede by exercising tactical restraint towards their civilians. The leverage exerted by Hezbollah, the Irgun, and the ANC certainly
didn't come from indiscriminate ISIS-type bloodletting. The leaders of these
groups studiously avoided this approach by instead targeting military personnel and other government instruments of the occupations. Lumping
these tactics together with terrorism leaves the false impression that civilian
targeting paid off.

Terrorist Leaders as Lumpers

Researchers aren't the only ones guilty of lumping when it comes to evaluating terrorism's alleged successes. There's considerable evidence that terrorists
themselves have overrated the political effectiveness of terrorism by drawing
false analogies from historically successful guerrilla campaigns. When terrorists look in the mirror, they often see guerrillas. Take Timothy McVeigh,
perpetrator of the worst domestic terrorist attack in American history.
McVeigh detonated a Ryder truck packed with explosives in front of the Alfred

P. Murrah building in downtown Oklahoma City on April 19, 1995. The vast majority of the 168 people killed were civilian, including nineteen children playing at the Kids Day Care Center.[40] One can never know for sure what inspires terrorists to take innocent life. But hours after the attack, McVeigh was arrested decked out in Revolution War garb. His T-shirt sported a picture of a tree dripping with blood and a quotation from Thomas Jefferson that read, "The tree of liberty must be refreshed from time to time with the blood of patriots and tyrants."[41] This wasn't just a random T-shirt he wore for the fit. McVeigh also carried an envelope with Revolutionary War literature, including a bumper sticker with another Jefferson slogan, "When the government fears the people, there is liberty." McVeigh had hand-written on the envelope, "Maybe now, there will be liberty!" alongside a John Lock quotation asserting that man has a right to kill those who trammel on his liberty.[42] According to the veteran police captain, Robert Snow, domestic terrorists in the Patriot Movement like McVeigh believe their terrorist acts are following the successful model of the Revolutionary War, even though it was directed against British soldiers and not their children.[43]

The radical left-wing Weather Underground seemed to think their bombing of American cities in the early 1970s would replicate the political triumphs of guerrilla campaigns in the Third World. Larry Grathwohl, the FBI informant who infiltrated the group, says members used to go around saying: "We can win. The same way the Viet Cong are beating back a half million U.S. troops" in the jungle of Vietnam.[44] The Vietnamese Communist revolutionary leader, Hồ Chí Minh, was the "patron saint" of the Weathermen.[45] In their communications to the underground, leaders told members that the violence in American cities was based on "the classic strategy of the Viet Cong," notwithstanding the obvious differences in target selection.[46] In her autobiography, the Weather Underground leader, Susan Stern, confirms that she believed her tactics would work because the Vietcong offered "a history for us to follow."[47] Stern acknowledges that members "thought of themselves as revolutionaries," indeed as the "Americong."[48] Laura Browder points out in the prologue to her autobiography that the problem for the Weather Underground was that "While their strategy was certainly theoretically grounded in their desire to engage in guerrilla struggle...the theory that they embraced did not lead them to effective forms of activism."[49] The Weather Underground seized on the Vietcong as a model to end oppression, even though its coercive success was against American troops rather than civilians.[50] Beyond the Weather Underground, New Left terrorist groups like Action Directe, Japanese Red Army, Red Army Faction, and Red Brigades sprang up throughout Asia, Europe, Latin America, and the United States, claiming inspiration in the political successes of the anti-colonial campaigns, which triumphed through the attrition of national militaries.[51]

The Malayan National Liberation Army (MNLA) also overrated the strategic value of terrorism by lumping together alternative tactical achievements. Lucian Pye conducted extensive interviews with the members and concluded that they sought to emulate victorious guerrilla campaigns, especially Mao Tse-Tung's ending of imperialism in China. As Pye notes, however, the MNLA didn't follow his famous dictum to spare civilians: "Although the party steadfastly continued to use the military terminology of guerrilla warfare to describe its activities…it was practicing terrorism, not warfare…The MNLA would strike against civilians…but this was terrorism and not guerrilla warfare."[52] More recently, Palestinian terrorists have overrated the political value of attacking Israeli civilians by drawing false analogies with successful guerrilla campaigns against military targets, especially Hezbollah's victory in coercing the Israel Defense Forces out of southern Lebanon in 2000. The leader of Palestinian Islamic Jihad explained the strategic rationale for blowing up Israeli civilians in buses, cafes, markets, and restaurants during the Second Intifada: "The shameful defeat that Israel suffered in southern Lebanon and which caused its army to flee in terror was not made on the negotiations table but on the battlefield and through jihad and martyrdom, which achieved a great victory for the Islamic resistance and Lebanese people. We would not exaggerate if we said that the chances of achieving victory in Palestine are greater than in Lebanon."[53] The Hamas leadership offered the same faulty rationale for killing Israeli civilians: "The Zionist enemy only understands the language of jihad, resistance, and martyrdom; that was the language that led to its blatant defeat in South Lebanon and it will be the language that will defeat it on the land of Palestine."[54] Anecdotal evidence thus abounds that researchers and terrorists alike have overrated the political value of terrorism by fixating on salient asymmetric victories that weren't even directed against civilians.

To examine the prevalence of this lumping phenomenon more systematically, I conducted a content analysis of the historical asymmetric campaigns that Osama Bin Laden invoked to justify the strategic logic behind 9/11. This content is made up of Bin Laden's translated statements issued in public and private between 1994 and 2004.[55] Collated by the Foreign Broadcast Information Service, this self-contained compilation of ninety-eight interviews, correspondences, and fatwas are thought to provide reliable insight into his strategic mindset during that crucial period.[56] If the analysis reveals that Bin Laden tended to invoke historically successful guerrilla campaigns to justify the prospects of 9/11 working then such evidence would strengthen my impression that lumping accounts for a sizable portion of alleged "terrorism" success stories. A potential limitation to this approach is that terrorist leaders have an interest in mobilizing constituents, so Bin Laden conceivably invoked successful guerrilla campaigns knowing full well that they're not actually analogous to blowing up the World Trade Center. As the China historian

John Fairbank remarked, history is a "grab-bag" from which each advocate pulls out a "lesson" to advance his agenda.[57] This is indeed a concern, but not a terminal one. Psychological research demonstrates that there's generally surprisingly little distance between discourse and reasoning.[58] To sidestep this potential methodological concern, though, the content analyzed includes private correspondence that wasn't intended for public consumption.[59]

Figure 3.1 reveals that all sixty-five of the asymmetric campaigns invoked are guerrilla—not terrorist—in the sense that they targeted the government's troops as opposed to its civilians. Figure 3.2 illustrates the campaign breakdown. Based on his public and private statements, Bin Laden appears to have overestimated the strategic value of 9/11 by drawing false analogies from the

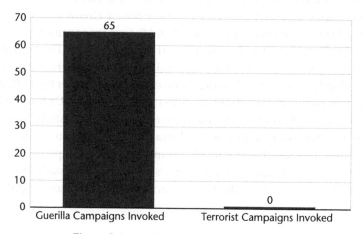

Figure 3.1. Analogies to Predict 9/11 Success

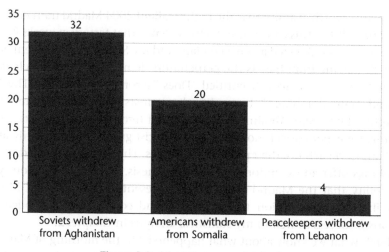

Figure 3.2. Most Common Analogies

political triumphs of three guerrilla successes in particular—the campaigns in the 1980s and 1990s that forced the peacekeepers from Lebanon, the Americans from Somalia, and the Soviets from Afghanistan. As we've seen, the U.S. and France withdrew from Lebanon after Hezbollah targeted their troops, not civilians. American forces left Somalia in 1994 after Mogadishu-based militia loyal to the warlord Mohamed Farah Aideed shot down two U.S. military helicopters commemorated in the *Black Hawk Down* movie. And the mujahedeen coerced the Soviets into leaving Afghanistan in 1988 by targeting the Red Army. Arguably, none of these asymmetric successes was due to terrorism as the violence was directed against occupying troops. Nonetheless, Bin Laden seemed sure terrorism would work because of these successful guerrilla campaigns, especially the one against the Red Army. The Al Qaeda leader reasoned incorrectly, "If the Soviet Union can be destroyed, the United States can also be beheaded," and "so we are continuing this policy in America."[60] In fact, Bin Laden thought that terrorizing the United States into submission "would be easier, God willing, than the earlier defeat of the Soviet Union" because "the Americans are cowards," a "paper tiger" that "after a few blows ran in defeat" from Lebanon and Somalia.[61] Unfortunately for him, however, "Lebanon and Somalia were the wrong yardsticks by which to measure the American response," as Fawaz Gerges notes in his book on Al Qaeda.[62] We thus see that the most commonly cited examples of "terrorism success stories" by researchers and terrorists alike were actually not cases of terrorism at all if we restrict the tactic to anti-civilian violence as splitters do.

2004 Madrid Train Bombing

A final case worth examining is the highly salient 2004 Madrid train bombing. Scholars point to this case as ipso facto evidence that terrorism works because the violence was directed against civilians and the Spanish government ended up withdrawing from Iraq as the perpetrators demanded. In an article in the journal *International Security* entitled "Does Terrorism Ever Work? The 2004 Madrid Train Bombings," the political scientists, William Rose and Rysia Murphy, state: "Max Abrahms' argument that terrorism rarely works is compelling. He is not correct, however, that terrorist groups that primarily attack civilians never achieve their political objectives. The March 2004 Madrid train bombings offer an exception to Abrahms's thesis."[63] Let's take seriously the possibility that the Madrid train attack represents what Rose and Murphy describe as an "exception to the rule."[64] Indeed, this case is coded as a terrorism success as well in my statistical analysis in Chapter 2.

Here's what we know about what happened. On the morning of March 11, 2004, Spain was the site of the most lethal terrorist attack in its history and the deadliest in Europe since 1988, when Libyan nationals blew up Pan Am Flight 103 over Lockerbie, Scotland. There's no question the target was civilian in nature. Thirteen bombs ripped through four commuter trains full of passengers, killing 191 people and injuring almost two thousand.[65] The attacks were directed by an Al Qaeda-inspired terrorist cell that came to be known as the 11-M network to mark the date of the tragedy. The primary political objective of the perpetrators may have been to compel Spain to end its military support for the U.S.-led occupations in Afghanistan and Iraq. In national elections just three days after the attack, Spanish voters defied earlier polls by choosing the anti-Iraq war candidate José Luis Rodríguez Zapatero over the incumbent, José María Aznar. As Rose and Murphy tell it, the "surprise winner" bombed to power was the Socialist Party, which during the campaign had called for removing Spanish troops from Iraq.[66] The troops were then promptly withdrawn in an apparently vivid demonstration of terrorism's strategic value. Incidentally, Aznar himself promotes this version of history that he didn't stand a chance after the train attacks.[67]

But even in this so-called "exception to rule" it's not even clear whether the train attacks resulted in the Zapatero victory. Indeed, he might have won the election and then altered Spanish policy in the absence of the attacks. The "surprise" defeat of Aznar was not all that surprising. In the days preceding the attack, Aznar held a narrow lead in some polls, but the differences between the candidates' voter estimates usually fell within the margin of error. By early March, the gap between the two candidates had closed; some surveys put Aznar ahead by a single point, while others had Zapatero winning by a razor-thin margin.[68] In a case study of the 2004 election, the political scientists Ignacio Lago and José Ramón Montero conclude: "It must be remembered that if the attacks had not taken place, either the PP [Aznar's Popular Party] or the PSOE [Zapatero's Socialist Party] could have won the election: only days before 11-M, the polls pointed to a technical tie."[69] In the lead-up to the attack, the majority of Spaniards believed the country needed a "change of government," a large percentage of the electorate was undecided, and, in Spain, undecided voters tend to break for left-wing candidates such as Zapatero.[70] Post-election returns confirm that Aznar didn't lose electoral support after the attack; as expected, the undecided voters simply gravitated toward the left-leaning candidates.[71] The extent of electoral change or inter-electoral volatility was typical for Spanish national elections.[72] The claim that 11-M successfully coerced Spain into withdrawing from Iraq is based on the counterfactual argument that Zapatero would have lost the election without the attack, which is unclear from the polling data.

Beyond the polling data, Spain-watchers are doubtful that undecided voters chose Zapatero because of the civilian costs to maintaining troops in Iraq. Instead, Aznar seems to have compromised his electoral viability by blaming the bombings on ETA, which was quickly exposed as a lie. Philip Gordon, the former director for European affairs at the National Security Council, testified to the U.S. Senate Foreign Relations Committee: "The [Aznar] government appears to have paid more of a price for misleading the public than for its policy on Iraq."[73] In her study on the election, Georgina Blakeley also concluded: "The point, therefore, is not that the bombings affected the general election, but rather, that the government's handling of the bombings had such profound consequences."[74] The *BBC* likewise reported: "It is sometimes wrongly claimed that the bombings themselves led directly to the defeat of the Conservative government and its replacement just days later by the Socialists. In fact, it was the perception that the government was misleading the public about who was responsible that did [the] most damage."[75] Journalists for *PBS News, Le Monde*, and Spanish television networks reached the same conclusion.[76] Although touted as clear-cut evidence of terrorism's political effectiveness, the outcome of the Madrid train attacks was thus anomalous, ambiguous, and idiosyncratic. The target of Spanish train passengers was unquestionably civilian in nature. Yet the subsequent withdrawal from Iraq depended on "uncommon conditions" that were "probably quite rare and possibly unique" as even Rose and Murphy admit.[77]

Does Terrorism Ever Work?

Revisiting the most celebrated "terrorist" victories underscores their rarity, particularly for splitters who count only violence directed against civilians. The point of this chapter isn't that terrorism has *never* paid off politically. Of course, we would fully expect some groups to have achieved their demands after using indiscriminate violence in the extensive history of asymmetric conflict. This admission doesn't invalidate the causal relationship between civilian attacks and political failure any more than my grandmother's long lifespan invalidates the scientific link between cigarette smoking and premature death. Exceptions are bound to pop up, but rebels would be wise to take it easy on the smoking and terrorism if they want to enjoy the spoils of victory into old age.

My research offers strong empirical support for those who have long questioned the political value of terrorism. In the 1970s, Walter Laqueur published "The Futility of Terrorism" in which he claimed that practitioners seldom seem to achieve their strategic demands.[78] In the 1980s, a RAND study observed: "Terrorists have been unable to translate the consequences of terrorism into concrete political gains...In that sense terrorism has failed. It is a fundamental failure."[79] Crenshaw made a similar observation around this time: "Few [terrorist] organizations actually attain the long-term ideological objectives they claim to seek, and therefore one must conclude that terrorism is objectively a failure."[80] In the 1990s, the Nobel Laureate Thomas Schelling noted, "Terrorism almost never appears to accomplish anything politically significant."[81] In Chapter 3, I show why.

Notes

1. Abrahms 2010.
2. Wilkinson 1986, 59.
3. Ibid. X.
4. Pape 2003.
5. Goodwin 2006, 317.
6. Levitt 2013.
7. Nance 2016, 313.
8. Kydd and Walter 2006, 49.
9. Gray 2009.
10. Wright 2006.
11. Gleis and Berti 2012, 80.
12. Norton 2007, 86.
13. "PM Says Israel Pre-Planned War," March 8, 2007.
14. Leung, 18 April 2003.
15. English 2016, 151.
16. Enders and Sandler 2012, 16.
17. Boot, 2013, 325.
18. Laqueur 2016, 118.
19. Hoffman 1998, 51.
20. Cronin 2011, 84–5.
21. Hoffman 2016, 298.
22. Hoffman 1998, 51.
23. Begin 1978, 213.
24. Hoffman 2016, 143, 204, 300, 304.
25. Cronin 2011, 84–5.
26. Hoffman 2016, 243.
27. Dershowitz 2002, 7.
28. Bryant 2003, 240.
29. Richardson 2007, 10.
30. Goodwin 2006, 235.
31. Mandela 1964.
32. Kathrada 2004, 105.
33. Turok 2003, 280.
34. Ibid. 284.
35. Battersby 1988.
36. O'Malley 2007, 207–8.

37. Mandela 2013, 240.
38. Lewin 2011, 116; O'Malley 2007, 233.
39. For a similar reading of the ANC, see Cronin 20011, 92.
40. Romano, February 27, 1997.
41. Fleeman, May 20, 1997.
42. Michel and Herbeck 2001, 228.
43. Snow 1999, 28.
44. Grathwohl and Reagan 1976, 100.
45. Ibid. 180–1.
46. Ibid.
47. Stern 2007, 201.
48. Ibid. 102, 192.
49. Ibid. xvii.
50. Laqueur 2016, 82.
51. See Wilkinson 1986.
52. Pye 1956, 95, 97–8.
53. Hoffman 2016, 154.
54. Ibid.
55. U.S. State Department 2004.
56. Ibid. 121.
57. Hoffmann 1968, 135.
58. Reiter 1996.
59. On the value of analyzing both public and private statements, see Axelrod 1976, 255.
60. U.S. State Department 2004, 15.
61. Ibid. 44, 49, 97, 235.
62. Gerges 2014, 92.
63. Rose, Murphy, and Abrahms 2007, 185.
64. Ibid. 188.
65. Chari 2004, 954.
66. Rose, Murphy, and Abrahms 2007, 187.
67. Aznar 2008. Anzar was the featured speaker at this event hosted by Stanford Law School.
68. Lago and Montero 2006.
69. Ibid. 17.
70. Blakeley 2006, 339.
71. Rigo 2005, 613.
72. Biezen 2005, 107.
73. Gordon 2004, 2.
74. Blakeley 2006, 342.
75. Ibid.
76. Suarez et al., March 16, 2004.
77. Rose, Murphy, and Abrahms 2007, 185, 188.
78. Laqueur 1976.
79. Cordes et al. 1984, 49.
80. Crenshaw 1987, 15.
81. Schelling 1991, 20.

4

Correspondence of Means and Ends Bias

Why does terrorism impede militant groups from achieving their political demands? This chapter proposes and tests an original psychological explanation to account for why target countries are so reluctant to grant concessions when groups attack their civilians. To understand why attacking civilians dissuades governments from concessions, it's useful to step back and consider how terrorism is supposed work in the first place. In theory, terrorism functions as a political communication strategy. The violence amplifies the demands of the perpetrators and signals the costs of noncompliance. Terrorism supposedly pressures concessions when the expected cost of compliance is lower than the anticipated cost from the violence. In other words, terrorism succeeds when the government determines that it's better off granting the demands of the perpetrators than incurring their continued wrath.[1] In reality, though, terrorism struggles to coerce concessions precisely because it's a flawed communication strategy.

A Flawed Communication Strategy

It's widely believed that violence helps international actors to communicate their grievances, largely by attracting media attention.[2] The adage "If it bleeds it leads" captures this line of thinking. The political scientist, Thomas Rochon, says "militancy" is what's "newsworthy" in European protests.[3] In his study of American protest groups, the sociologist, William Gamson, remarked, "To keep media attention, challengers are tempted to use ever-more extravagant and dramatic actions."[4] Some leaders admit to adopting violence for the media attention. The head of the United Red Army, an obscure offshoot of the Japanese Red Army, acknowledged: "There is no other way for us. Violent actions...are shocking. We want to shock people everywhere...It is our way of communicating with the people."[5] Similarly, Bin Laden and his deputy, Ayman al-Zawahiri, described Al Qaeda violence as a "message with no words" which is "the only language understood by the West."[6] And yet the sociologists, Donatella Della Porta and Mario Diani, remind us: "It is the content of the message transmitted as well as the quantity of publicity received which is important for the social movement."[7] By definition, terrorism gets noticed; otherwise, it wouldn't terrorize. But does the violence help the perpetrators to convey their grievances?

The evidence is less than convincing. Historical accounts point out that the Weather Underground could "bomb their names on to the front pages, but they could do next to nothing to make sure that the message intended by their bombings was also the message transmitted."[8] Most Italians in the 1970s viewed the Red Brigades as "common criminals" rather than left-wing revolutionaries.[9] For West European terrorists, "The media generally present[ed] images of their protest without any elaboration of the substantive issues involved."[10] Other research on European terrorist groups in the 1980s likewise determined that "although terrorism is often described as a form of communication, terrorists are rather poor communicators," as "the violence of terrorism is rarely understood by the public."[11] The sociologist, Charles Tilly, observed that American journalists didn't grasp the political purpose of Chechen hostage-taking in the 1990s beyond "senseless acts" of violence.[12] In 2003, the Ugandan newspaper, *New Vision*, claimed that while the Lord's Resistance Army killed countless civilians, the group has "absolutely no political ideology" and lacks even "a veneer of ideology."[13] A study on IRA and Iranian-sponsored terrorism found that "the terrorist motives and goals were ignored."[14] Instead, the media depicted their violence as "senseless bestiality."[15] In the 1980s, Michael Kelly and Thomas Mitchell did a content analysis of 158 terrorist incidents covered in the *New York Times* and the *Times of London*. The terrorism seemed to "sap . . . its political content" as "less than 10 percent of the coverage in either newspaper dealt in even the most superficial way with the grievances of the terrorists."[16] Leaders of aggrieved groups often lament that the violence didn't have the desired communicative effect. Mac Maharaj of the ANC, for example, worried that blowing up the Wimpy and Magoo bars in the 1980s only "reinforce[d] whites' belief that blacks were intent on murdering whites simply because they were white."[17] Together, these observations provide a clue that perhaps terrorism is an ineffective coercive tactic because it's a faulty communication strategy.

We know that victims of assault, battery, robbery, and other violent crimes systematically misperceive the motives of the perpetrators. The social psychologist, Roy Baumeister, discovered that violent crime victims suffer from what he calls "The Myth of Pure Evil." Violent crimes are almost always instrumental in nature; that is, the perpetrator commits them as a means to an end rather than as an end in itself. For this reason, he'll usually desist upon attaining his demands or finding an easier way to achieve them.[18] But this isn't how victims interpret the violence at all. Two versions prevail in victim accounts. In one, the victim stresses that the perpetrator had no reason at all to do what he did. This account emphasizes "the utter outrageousness, the sheer incomprehensibility of the perpetrator's action" which seems "arbitrary, random, gratuitous." The other version presents the perpetrator as a sadistic monster who is "deliberately malicious" driven by the "desire to do harm as an end in itself" just for "the pleasure of doing so."[19] Children's cartoons play into the Myth of Pure Evil. The psychologists, Petra Hesse and John Mack, analyzed the villains in the eight highest-rated cartoons from the late 1980s. The cartoon villains were cast as having no apparent reason for their attacks other than the joy of making innocent people suffer. Evil was done for evil's sake.[20]

Correspondence of Means and Ends Bias

Building on Baumeister, I submit that the purpose of terrorist violence is generally misunderstood by the target country. In fact, the act of terrorism itself impedes perpetrators from conveying their preferences, undermining the communicative and thus political value. Terrorism is a faulty political communication strategy because of a newfound cognitive heuristic in international affairs that I've coined as the Correspondence of Means and Ends bias.[21] Social scientists have previously uncovered other cognitive heuristics that outperform assumptions of perfect rationality.[22] The proposed bias is that citizens of target countries don't locate the goals of terrorists in their demands, so the violence fails to amplify them. Rather, people tend to infer the extremeness of the perpetrators' goals directly from the extremeness of their tactics against them. As its name suggests, the bias posits that citizens draw a direct correspondence between the extreme means of their adversary and its extreme ends. Notwithstanding the nature of their actual demands, terrorists are thus seen as harboring radical political preferences by dint of their tactics. In the parlance of economics, citizens of target countries view the extreme means of terrorism as endogenous to its intended ends. This cognitive bias kills off any incentive for target countries to negotiate because terrorists are perceived as unappeasable extremists even when their demands are surprisingly moderate. Citizens of target countries tend to dismiss them as insincere, even irrelevant because the terrorists are deemed too radical to realistically be appeased. Because of this human tendency to conflate the extreme means of terrorists with their presumed ends, tactical escalation discredits their vow to remove the pain even if government concessions were to be made. Seen as implacable foes regardless of their demands, terrorists are written off as unreliable negotiating partners better suited for target practice than the bargaining table.[23]

Let's test the Correspondence of Means and Ends bias. For starters, the extremeness of both an actor's means (tactics) and ends (political preferences) can be ranked on stylized continua. Within international relations, the extremeness of an actor's tactics is ranked in terms of the pain or physical costs to the population. That is, killing civilians is regarded as an extreme method compared to leaving the population unharmed.[24] The extremeness of political preferences is ranked in terms of the costs of government accommodation. At one end of the spectrum are moderates, who seek money, prisoner releases, or other tangible resources that are relatively inexpensive to relinquish. At the opposite end are extremists, who are motivated to harm the citizens of the target country for whatever reason, be it their ethnicity, religion, or way of life.[25] If the Correspondence of Means and Ends bias is valid, we should therefore find that people are apt to conclude that an adversary harbors extreme political preferences when he employs extreme tactics. Specifically, citizens of target countries should be more inclined to believe that a group is bent on harming them when it escalates with terrorism—even in cases in which the perpetrators promise to demobilize in exchange for prisoners, money, or other moderate demands.

To test this cognitive bias, I've conducted what's known as an experiment embedded in a survey. The survey research firm, YouGov, fielded my experiment over the internet on a large, national sample of voting-age American citizens. By using random digit dialing to recruit participants and providing free internet access to households that lack it, YouGov administers survey instruments online to representative national samples. This technique has become standard in survey experiments because the samples generated tend to surpass in quality those from conventional telephone polling.[26]

All subjects were presented with a simple vignette of an unidentifiable group issuing a traditionally moderate preference through the American media—the release of its imprisoned leaders from U.S. custody in exchange for permanently demobilizing. Subjects were randomly assigned, however, to two conditions that differed along a tactical dimension. In the control condition, the group surrounds a bunch of American civilians, takes them hostage, but doesn't physically harm anyone in the course of the confrontation. The same information was presented in the treatment condition, except the group escalates tactically by killing the civilians in its custody. To minimize framing issues, I paid attention to the formal aspects of the survey instrument by avoiding any derivatives of the word terror or any other emotive labels to describe either the coercive acts or the actors themselves. The two conditions were thus duplicates, except in the terrorism treatment the moderate group adopts a more extreme method by killing the civilians instead of releasing them unharmed.

Subjects in both conditions were then presented with a series of identical multiple choice and ordinal scale questions designed to assess both directly and indirectly the perceived extremeness of the group's goals. Specifically, all subjects were asked the following set of questions: (1) to evaluate whether the group is motivated to achieve its demand of freeing the imprisoned leaders in U.S. custody or to harm Americans out of hatred towards them; (2) to rate the group's preferences from 1 to 7 along this continuum;[27] (3) to judge whether the group would demobilize upon achieving its demand to free the imprisoned leaders; (4) to appraise whether the group would gain satisfaction from Americans physically harmed in an unrelated incident that wouldn't contribute to winning back the imprisoned leaders in U.S. custody; and (5) to ascertain whether the group would continue to engage in the same actions against Americans even after discovering a less extreme method that promised to free the imprisoned leaders.

Following convention in experimental research, I applied a two-tailed difference of means test to determine whether the tactical manipulation alone yields significant variation in the perceived extremeness of the self-described moderate group's preferences. Answers to each of the five questions are strongly confirming of the Correspondence of Means and Ends bias and statistically significant at the .01 level or better. Table A.5 presents the results which are illustrated in Figure 4.1. Compared to subjects in the control condition in which no civilians were physically harmed, those exposed to the terrorism treatment were on average: (1) 27 percent more likely to believe the group is motivated not to free the imprisoned leaders in U.S. custody, but to harm Americans out of hatred towards them; (2) 20 percent more likely to rate the group's preferences as the most extreme on a standard 7-point ordinal scale;[28] (3) 23 percent more likely to believe the group wouldn't demobilize upon achieving its demand to free the imprisoned leaders; (4) 33 percent more likely to believe the group would gain satisfaction from Americans physically harmed in an unrelated incident that wouldn't contribute to winning back the imprisoned leaders; and (5) 22 percent more likely to believe the group would continue to engage in the same actions against Americans even after discovering a less extreme method to free its leaders from U.S. custody.

For experimentalists in psychology or behavioral science, the quality of a causal mechanism depends less on the number of confirming cases than on the theoretical construct and its predictive power.[29] As a robustness check, though, I also tested the mechanism with another vignette, where this time the perpetrators demanded that the government cede not prisoners but money in exchange for permanently demobilizing. These two demands are by far the most commonly found ones in contemporary hostage-taking situations according to the ITERATE dataset.[30] Once again, subjects were exposed to the same information about the hostage-takers, except, in the terrorism treatment, respondents were told that the group escalates tactically by killing the captives instead of releasing them unscathed. Subjects in both conditions were presented with the same set of questions to further assess

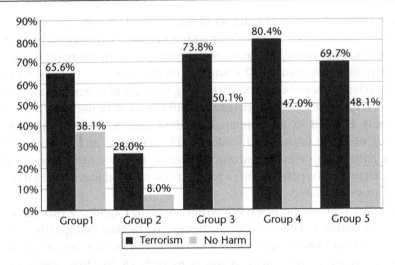

Group 1 Group hates Americans and wants to harm them

Group 2 Group is most extreme on 7-point scale

Group 3 Group would not demobilize upon achieving demand

Group 4 Group would gain satisfaction from Americans harmed in unrelated incident

Group 5 Group would continue same actions upon discovering less extreme
tactic to achieve demand

Figure 4.1. Terrorism Treatment Versus No Control

whether the extremeness of tactics employed by the perpetrators informs perceptions of their preferences independently of the actual demands. Table A.6 in the Appendix presents these results. Across questions, those exposed to the terrorism treatment were again significantly more likely to conclude that the perpetrators are evidently motivated to harm the population irrespective of whether the moderate demand was granted.

The Nobel laureate, Thomas Schelling, was the first to formally theorize that in international politics the type of violence used is itself part of the negotiation process. He said that hostage-taking represents the purest form of coercive bargaining because the perpetrators can pressure compliance by ratcheting up the credibility of their threats with pain.[31] This simple experiment elucidates how pain may strengthen the credibility of a threat but weaken the credibility of the promise to remove it, negating the logic of negotiations. This experiment helps to explain my finding of why target countries are so reluctant to compromise with terrorists. As the Correspondence of Means and Ends bias predicts, citizens of target countries seem to infer that terrorists harbor extreme, unappeasable political preferences due to their tactics, notwithstanding their actual demands.

The remainder of this chapter demonstrates the external validity of this theory with three case studies: the responses of Russia to the September 1999 apartment bombings, the United States to the September 11 attacks, and Israel to Palestinian terrorism in the First Intifada. These cases offer additional evidence that (1) target countries infer groups have extreme political objectives when they commit terrorism regardless of their actual demands, and (2) this belief that the perpetrators are political extremists dissuades concessions. More specifically, the cases illustrate that countries believe their populations are attacked not because the terrorists are protesting specific grievances like territorial occupation or poverty. Rather, target countries infer from the tactic of terrorism itself—the deaths of innocent citizens, mass fear, loss of confidence in the government to offer protection, economic contraction, and the erosion of civil liberties—the objectives of the perpetrators. In short, target countries view the negative consequences of terrorist attacks on their societies as proof the terrorists want them destroyed. Target countries are understandably skeptical that making concessions will placate adversaries believed to be motivated by such extreme ends. As a consequence, terrorists struggle to coerce government compliance even when their demands are surprisingly moderate.

Russia's Response to the 1999 Apartment Bombings

Russia's response to the three apartment bombings that killed 229 Russian civilians in September 1999 illustrates why groups are less likely to achieve their demands when they target civilians. Before news of the bombings reverberated throughout Russia, there was widespread agreement among Russians that Chechen objectives were limited to establishing an independent Chechen state. During this period, most Russians favored territorial compromise. After the apartment bombings, though, Russian society inferred from them that the presumed perpetrators evidently want to destroy Russia. Below I detail how this attitudinal shift from the bombings eroded Russian support for Chechen independence.

With the collapse of the Soviet Union in the late 1980s, several North Caucasian republics declared sovereignty. Chechnya's first president, Dzhokar Dudayev, took the additional step in 1991 of declaring independence. Federal forces invaded Chechnya in December 1994 to reestablish control in the breakaway republic. For the next twenty months, Russian federal forces battled Chechen guerrillas in an asymmetric war based in Chechnya. During this period of guerrilla warfare, Russians recognized that Chechen objectives were limited to self-determination. The Russia specialist, John Russell, noted at this time: "The Russian public [believed] that Chechens perceived their struggle as one of national liberation."[32] The Russia diplomat, Michael McFaul, similarly observed: "The Russian people believed that the [Chechen] rationale for this war was self-defense."[33] The Russian military shared this view that Chechen objectives were territorial, calling the first Chechen war the War for the Restoration of Territorial Integrity.[34]

During this period, most Russians were prepared to make major concessions over the status of Chechnya. When the war broke out, the Russian public and even the secret police perceived it as precipitate, believing diplomatic solutions hadn't been exhausted.[35] Boris Yeltsin's pro-war stance failed to gain traction. Top military commanders openly resigned and condemned the president for not pursuing negotiations.[36] From the onset of military operations until the cease-fire in August 1996, some 70 percent of Russians opposed the war.[37] Disdain for the war manifested itself most clearly in public attitudes toward Defense Minister Pavel Grachev. Opinion polls rated his approval at only 3 percent, just a few points lower than the Russian public's support for Yeltsin's handling of the Chechen problem in general.[38] By early 1996, domestic opposition to fighting the guerrillas imperiled Yeltsin's electoral prospects. *The Economist* predicted that February, "Mr. Yeltsin can scarcely afford to let the conflict drift violently on if he hopes to win a second term of office in June's presidential election."[39] Yeltsin folded to domestic pressure, calling an end to all troop operations in Chechnya, and the immediate commencement of negotiations with Dudayev over its future status. Yeltsin's approval rating shot up from 21 percent in February 1996 to 52 percent three months later. In May, Yeltsin admitted in an interview that he would lose the upcoming election if he didn't grant Chechnya de facto sovereignty. The Khasavyurt agreement of August 1996 committed the Russian Federation to relinquishing Chechnya by December 2001.[40] In the interim period, Chechnya would have a de facto state with free elections, a parliament, a president, and autonomy over its finances and resources.[41] Noting the generous terms of the accord, the *BBC* remarked that the status bestowed upon Chechnya "essentially signified its independence."[42] In Khasavyurt, Yeltsin formally acknowledged that Russians preferred territorial compromise to fighting the guerrillas.

But then Russia experienced its "first taste of modern-day international terrorism."[43] On September 8, 1999, a large bomb was detonated on the ground floor of an apartment building in southeast Moscow, killing ninety-four civilians. On September 13, another large bomb blew up an apartment building on the Kashmirskoye highway, killing 118 civilians. On September 16, a truck bomb exploded outside a nine-story apartment complex in the southern Russian city of Volgodonsk, killing seventeen civilians. The Kremlin quickly fingered the Chechens as the perpetrators.[44]

As the Correspondence of Means and Ends bias would predict, Russians responded to the terrorist attacks by fixating on them rather than Chechen policy demands. Russia watchers noted during the bombing campaign,

"Attention is being directed more at the actual perpetration of terrorist acts, their visible, external and horrific effects with perhaps not so much emphasis on the [stated] cause."[45] The terrorist violence altered perceptions of Chechen intentions. Polls show that following the terrorist acts only 15 percent of Russians believed the Chechens were fighting for independence. The Public Opinion Foundation reported: "This motive is mentioned less frequently now... [It] is gradually dying out."[46] As Russia analyst, Timothy Thomas, observed after the bombings, "The Chechens were no longer regarded as a small separatist people struggling to defend their territory."[47] Yeltsin's successor, Vladimir Putin, contributed to the view that the Chechens ceased to be motivated by the desire for national self-determination, declaring, "The question of Chechnya's dependence on, or independence from, Russia is of absolutely no fundamental importance anymore." After the terrorist attacks, he stopped referring to Chechens altogether, instead calling them "terrorists." The campaign for the "restoration of territorial integrity" became "the campaign against terrorism" and Russian counterattacks became "counterterrorist operations." Putin's focus on Chechen terrorism was most evident after September 11, 2001, when he told the United States that "we have a common foe" since the apartment bombings "bear the same signature" as the World Trade Center attacks.[48]

Following the bombings, Russians concluded that Chechen objectives had suddenly become more extreme. Polls show that Russians were now almost twice as likely to believe that Chechen motives were to "kill Russians," "bring Russia to its knees," "destabilize the situation in Russia," "destroy and frighten Russian society," and "bring chaos to Russian society" than to achieve "the independence of Chechnya."[49] Putin's public statements suggest that he too inferred the extremeness of Chechen objectives directly from the tactics, asserting that the perpetrators are evidently attacking Russia so it "goes up in flames."[50] This post-bombing belief that Chechen objectives had suddenly become so extreme was accompanied by an abrupt loss of interest in concessions and unprecedented support for war. The Public Opinion Foundation found that 71 percent of Russians had supported the idea of trading land for peace but came to believe "the Chechens are not trustworthy."[51] When Russians were asked to explain why they no longer trusted the Chechens to abide by a land-for-peace deal, the most common explanation was "because of the terrorist acts."[52] Whereas Russians had demanded Yeltsin's impeachment over the first Chechen conflict, after the apartment bombings they were "baying for blood."[53] In the first Chechen war, Russians favored by a two to one margin an independent Chechen state over battling the guerrillas in the breakaway republic; after the bombings, these numbers were reversed even when respondents were told that federal forces would "suffer heavy losses."[54] Popular support for war remained remarkably stable for many years after the bombings.[55]

In the mid-1990s, foreign jihadists began using Chechen territory as a safe haven. Chechen terrorists have indeed become more politically extreme over time. But the point is that Russian perceptions of Chechen aims changed profoundly as a direct result of the apartment bombings. The terrorism failed to highlight the costs of occupying Chechnya. As the Correspondence of Means and Ends bias would predict, Russians fixated on the bombings themselves and suddenly concluded that the suspected attackers evidently want Russia destroyed. Once Russians believed that the Chechen goal was no longer confined to achieving national self-determination, enthusiasm for compromise abruptly declined while support for a military solution soared.

U.S. Response to the September 11 Terrorist Attacks

The American response to the September 11 attacks further illustrates the difficulty groups face when they rely on terrorism as a communication strategy to amplify their demands. Instead of focusing on Al Qaeda's stated grievances after the attack, Americans fixated on the terrorism and inferred that the group was evidently motivated to harm American society as an end in itself. This perception that Al Qaeda was intent on destroying American society rather than achieving its given aims in the Muslim world impeded the group from achieving them.

Bin Laden's criticisms leading up to 9/11 didn't focus on hatred of American people, culture, or values.[56] As we saw in Chapter 1, he repeatedly stated that the purpose of 9/11 was to coerce the U.S. into ending its occupation of the Persian Gulf, relations with Israel and pro-Western Muslim rulers, and military operations that kill Muslims around the world. But Al Qaeda terrorism failed to broadcast these grievances or at least Americans didn't hear them. Ronald Steel noted in the *New Republic* that the American media focused on "what bin Laden attacked" and "the method of attack," but "what bin Laden had been saying about why he and his Al Qaeda forces were attacking was given short shrift."[57] The British journalist, Robert Fisk, likewise observed that Americans fixated on the terrorist carnage, but "not in a single press statement, press conference, or interview did a U.S. leader or diplomat explain why the enemies of America hate America."[58] Western journalists devoted generous coverage to the attacks, but only with the publication of Michael Scheuer's *Imperial Hubris* in 2004 did they regularly publish Al Qaeda communiqués.[59]

President George W. Bush seems to have deduced Al Qaeda's motives directly from the 9/11 attacks themselves. According to Bush, "We have seen the true nature of these terrorists in the nature of their attacks" rather than in their professed political agenda.[60] For Bush, September 11 demonstrated that the enemy "hates not our policies, but our existence."[61] The stated political ends of the terrorists were irrelevant. Any group that deliberately attacks American civilians is evidently motivated by the desire to destroy or harm them. When asked by a reporter in October 2001 if there was any direct connection between the September 11 attacks and the spate of anthrax letter attacks that followed, Bush replied: "I have no direct evidence but there are some links . . . both series of actions are motivated to disrupt Americans' way of life."[62] This interpretation of the unknown terrorist perpetrator is revealing; whoever it was must have sent the anthrax because he's a political extremist bent on harming American society. President Bush isn't the only U.S. president to infer the political objectives of terrorists from their attacks. After the 1993 World Trade Center attack, President Bill Clinton said the purpose was apparently to destroy American society.[63] The American public shared this interpretation of the terrorist motives. Polls conducted after September 11 show that most Americans believed that Al Qaeda wasn't responding to unpopular U.S. foreign policies despite its political demands. Only one in five respondents agreed with the statement that there was any way that "The United States has been unfair in its dealings with other countries that might have motivated the terrorist attacks."[64] In a separate poll, only 15 percent of Americans agreed that "American foreign policies are partly the reason" for Al Qaeda terrorism.[65] Americans were far more likely to tell pollsters that the U.S. was struck because of its "democracy," "freedom," "values," and "way of life" than policies in the Muslim world.[66]

Bin Laden and his lieutenants frequently complained that Americans failed to "understand" the "true reason" for the September 11 attacks. Instead of attacking because we hate you and your society, he said, the attacks are a response to the fact that "you spoil our security" and "attack us."[67] On multiple occasions, he warned that those who repeat this "lie" either suffer from "confusion" or are intentionally "misleading you."[68] The Correspondence of Means and Ends bias elucidates the failure of Al Qaeda's communication strategy. As the bias predicts, September 11 didn't amplify Al Qaeda's policy grievances. Instead, the American public and president focused on the terrorist attacks themselves and deduced from them that the terrorists must want to destroy Americans.[69] Admittedly, even if terrorism hadn't delegitimized Al Qaeda's policy demands, it's inconceivable the U.S. would have ever fully complied with them. But it's also doubtful Americans would have favored a counterterrorism strategy after September 11 that systematically exacerbated Bin Laden's stated grievances. As we saw in Chapter 1, the U.S. responded to 9/11 by increasing troop levels in the Persian Gulf fifteen-fold, strengthening military relations with pro-U.S. Muslim rulers, supporting counterterrorism missions that killed tens of thousands of Muslims around the world, and becoming an even less partial mediator in the Israeli-Palestinian conflict. Al Qaeda's post–September 11 policy failures are a testament, at least in part, to its flawed communication strategy.

Israel's Response to the First Intifada

The First Intifada may seem like an unlikely case study to illustrate the limitations of terrorism as a communication strategy. The mass uprising in the Gaza Strip and West Bank was a relatively moderate period in the history of Palestinian terrorism. The revolt from December 1987 to January 1991 killed only twenty Israeli civilians. Compared to the so-called Revolutionary Violence campaign of the 1970s and the outbreak of the Second Intifada in September 2000, the First Intifada was a peaceful interlude.[70] Facing relatively little competition from other Palestinian groups, the Palestine Liberation Organization co-opted the mass uprising in the late 1980s by recognizing the Israeli state within its pre-1967 borders and formally renouncing terrorism. Despite the unusually moderate tactics and objectives of the Intifada, Israelis interpreted the limited use of terrorism as evidence that the Palestinians wanted them destroyed, undermining support for territorial concessions.

As the conflict researcher, Edy Kaufman, has noted, "The primary purpose of the First Intifada was to communicate to Israelis the need to end the occupation of the territories."[71] But terrorist acts, even in small numbers, interfered with the message. Only 15 percent of Palestinian demonstrations throughout the Intifada were violent.[72] Yet 80 percent of Israelis thought the means employed by the Palestinians to protest Israeli rule were "mainly violent." The Palestinian attacks consisted almost entirely of rock throwing against the Israel Defense Forces in the territories, with few incidences of terrorism inside the Green Line. But 93 percent of Israelis felt the Intifada was directed "both towards civilians and towards the army."[73] Notwithstanding the Intifada's tactical restraint, Israelis appear to have fixated on the intermittent attacks against civilians.

The Louis Guttman Israel Institute of Applied Social Research conducted a series of polls in December 1990 to assess the Israeli public's views of Palestinian objectives in the First Intifada. As the Correspondence of Means and Ends bias predicts, 85 percent of the respondents believed the purpose was to "cause damage and injury" while only 15 percent believed the goal was to "express protest." Polls show that 66 percent of Israelis believed the Intifada was directed against "the existence of the state of Israel," while only 34 percent believed the purpose was to liberate the West Bank and Gaza Strip.[74] The disconnect between the PLO's policy demands and Israeli perceptions of its objectives may be partially explained by inconsistent rhetoric on the part of Palestinian leaders about the aims of the Intifada. But the attacks themselves also informed Israeli perceptions of Palestinian objectives. In a fascinating study based on the polling data contained in the Guttman report, Kaufman observed that the respondents who perceived Palestinian tactics as "mainly violent" were more likely to believe that the Palestinian goal was to "destroy Israel." Conversely, the more Israelis perceived Palestinian tactics as nonviolent, the more they believed the goal was to liberate the territories. This correlation between the perceived extremeness of Palestinian means and ends existed independent of the respondents' political affiliation, suggesting that this association wasn't a function of their preexisting ideological stances.[75] Not surprisingly, Israelis were twice as likely to believe "less in the idea of peace" than before the Intifada.[76] Because the majority of Israelis regarded the Intifada as a protracted terrorist campaign, and Israelis inferred from Palestinian terrorism their intentions of wanting to destroy Israel, the Intifada undermined Israeli confidence in the Palestinians as a credible partner for peace.

In the early 1990s, Israeli Prime Ministers Yitzhak Shamir and Yitzhak Rabin came under increased pressure to trade "land for peace" with the Palestinians. The sources of pressure were two-fold. First, President George H.W. Bush was determined to improve U.S.–Arab relations after Israel had lost its strategic utility as a Cold War satellite, so he "forced the Israelis to the negotiating table" by linking U.S. financial assistance to Shamir's participation in the Madrid peace conference in October 1991. Second, Israeli military strategists recognized that the Jewish state faced a long-term demographic problem in occupying a growing Palestinian population.[77] In September 1993, Israel consented to the land-for-peace formula outlined in the Oslo accords, but the pattern persisted: Although Palestinian terrorism demonstrated to Israel the costs of the occupation, it undercut Israeli confidence in the Palestinians as a credible partner for peace, reducing support for making territorial concessions.[78] Throughout the 1990s, the Jaffee Center for Strategic Studies periodically polled Israeli respondents on their perceptions of Palestinian aspirations. The "dominant" response was that the Palestinians wanted to "conquer Israel" and "destroy a large portion of the Jewish population," a position that peaked during heightened levels of terrorist activity.[79] This perception that the Palestinians hold extreme political aspirations has been a recurrent impediment to Israeli territorial concessions. Palestinian violence has created pressure on Israel to change the political status quo. Yet the terrorism has simultaneously convinced Israelis that the Palestinians aren't committed to a two-state solution, which has eroded support for making the concessions necessary to achieving it.

Scope Conditions

In the 1970s, members of the far-left Turkish People's Liberation Army kidnapped a teenage girl and got in a shootout with police. One of the terrorists shouted to the watching crowd, "We are doing this for you!" The crowd had no idea what he was talking about and then tried to lynch him.[80] In theory, terrorism shouldn't work this way. The violence supposedly functions as an effective political communication strategy that broadcasts the demands of the perpetrator, signals the costs of ignoring them, and ultimately induces concessions to stave off future pain. In reality, terrorism is a losing instrument of coercion precisely because it's an inherently flawed communication strategy. In this chapter, we saw both experimental and observational evidence for a newfound cognitive heuristic in international affairs that I call the Correspondence of Means and Ends bias. The bias is that people generally don't locate the preferences of terrorist adversaries in their political platform. Rather, citizens of target countries tend to infer the extremeness of their political ends directly from their means. For this reason, target countries tend to regard terrorists as unappeasable extremists even when their demands are surprisingly moderate, thereby eroding support for concessions.

Like all theories, this one has scope conditions. The Correspondence of Means and Ends bias is a cognitive heuristic, so it takes effect under conditions of uncertainty. Heuristics are imperfect substitutes for complete information as the psychologists of judgment, Daniel Kahneman and Amos Tversky, famously revealed.[81] We would therefore expect to observe the bias when the victim is relatively ignorant about the perpetrator's goals beyond his tactical choices. The 9/11 attack is a prime example. Even the White House knew next to nothing about Bin Laden's stated aims before the strikes.[82] So, it's unsurprising that Americans would base their understanding of Al Qaeda's goals on the violent acts themselves. By contrast, Spanish opinion of Al Qaeda's political objectives were already firmly established before the March 2004 Madrid train attacks. Polls show that 90 percent of the public disagreed with Aznar's position that participating in the Iraq war made Spain safer from terrorism. This Spanish perception had both breadth and depth, as reflected in some of the largest national protests in its history against the Iraqi occupation. Whereas news of the Chechnya occupation was largely withheld from the Russian public until it was targeted in September 1999, Spanish combat deaths in Iraq in August, October, and November 2003 were front-page news, reinforcing the view that the terrorists aimed to end the occupation rather than Spain's way of life.[83] Terrorists may stand a better chance of achieving their demands when they're already understood by the public. The Israeli case suggests, however, that even relatively minor amounts of terrorism in long-standing conflicts can distort perceptions of perpetrator goals. Chapter 4 indicates that attacking civilians not only reduces the odds of government concessions, but expedites the demise of militant groups, especially when the indiscriminate violence is perpetrated on a mass scale. Rebels should thus avoid terrorism even when they're not looking to bargain.

Notes

1. On this logic, see Lapan and Sandler 1993; Overgaard 1994.
2. McCarthy et al. 1996.
3. Rochon 1988, 102.
4. Gamson 1975, 166.
5. McKnight 1974, 168.
6. See, for example, "Interview with Osama Bin Laden," 10 May 1997.
7. Della Porta and Diani 2006, 180.
8. Lichbach 1998, 113.
9. Schmid and Graaf 1982, 111.
10. Della Porta 2006, 123.
11. Rochon 1988, 102.
12. Cordes et al. 1984, 1.
13. Tilly 2003, 237.
14. "Heed Final Call, LRA," October 10, 2003.
15. Hewitt 1993, 46–7.
16. Hewitt 1993, 52.
17. Kelly and Mitchell 1981, 287.
18. O'Malley 2007, 207–8.
19. Baumeister 1999, 101–3, 127.
20. Ibid. 42.
21. For more on this bias, see Abrahms 2013.
22. Kahneman and Tversky 1996.
23. This mechanism explains why terrorists suffer a credible commitment problem. On this problem, see also Findley and Young 2011.
24. DeNardo 1985, 190.
25. Marinov 2005.
26. Berinsky et al. 2012.
27. Respondents selected 1 if they believed the motive was based entirely on hatred towards Americans, 7 if they believed the motive was based entirely to achieve the moderate demand, and 4 if they believed both motives applied equally. The unlabeled values between these options allowed a more nuanced response. A response of "uncertain" was also an option for all subjects.
28. Subjects exposed to the terrorism treatment were also 13 percent less likely to rate the group's preferences as the least extreme ($p < 0.001$). The middling option was selected the fewest times, indicating that the rubric of preferences adequately captured respondent perceptions.
29. McDermott 2002.
30. ITERATE.
31. Schelling 1966, 192–4.
32. Russell 2002, 80.
33. McFaul 2000, 21.
34. Russell 2002, 77.
35. Bowker 2004, 469, 473.

36. Shevtsova 1999, 117.
37. Russell 2005, 105.
38. Shevtsova 1999.
39. "Russia and Chechnya," February 10, 1996.
40. Smith 2000, 5.
41. "Chechnya's Truce," June 10, 1996.
42. "Kremlin Deputy Chief," 4 October 2004.
43. Russell 2002, 77. The September 1999 apartment bombings were not the first terrorist attacks against Russian civilians, but there is general agreement that the bombings represented a watershed. Henceforth, Russian civilians would become a primary target.
44. Since the bombings, a conspiracy theory has circulated that the Federal Security Services framed the Chechens to build support for a counteroffensive. Less than 10 percent of Russians accept this theory as credible. The essential point is that the Russian public believed that the Chechens were responsible for the terrorist attacks, which negatively affected perceptions of their political motives.
45. Blandy 2003, 422.
46. "Chechen Labirinth," 2002.
47. Thomas 2000, 116.
48. Khalilov 2002, 411.
49. "The Bombing of Houses in Russian Towns," September 14, 2000.
50. Dlugy, December 29, 2004.
51. Klimova, November 21, 2002.
52. Ibid., January 30, 2003.
53. Herd 2000, 32.
54. "Chechnya—Trends 2000–2005," September 30, 2000.
55. Smith 2000, 6.
56. Al Qaeda almost always emphasized its hatred of U.S. foreign policies, not American values. See, for example, Abrahms 2006.
57. Steel, July 28, 1996.
58. Fisk, August 22, 1998.
59. Scheuer 2004.
60. "Speech by the President," November 6, 2001.
61. "Remarks by the President," November 10, 2001.
62. "Remarks by the President," October, 24 2001.
63. Clarke 2004, 129–30.
64. Seventy percent of Americans rejected the idea that "unfair" U.S. foreign policies contributed to the terrorist attacks.
65. Ipsos, September 21, 2001.
66. Harris Poll, September 24, 2001.
67. Scheuer 2004, 153.
68. Bin Laden, October 30, 2004.
69. Spiers, January 14, 2003.
70. B'Tselem.
71. Kaufman 1991, 4.

72. Sharp 1989, 7.
73. Kaufman 1991, 4.
74. Guttman 1990.
75. Kaufman 1991, 4.
76. *Jerusalem Post*, August 27, 1988.
77. Shlaim 1992, 4.
78. The limited objectives of the First Intifada shouldn't be confused with the maximalist objectives of some Palestinian terrorist organizations.
79. Arian 2000, 14.
80. Schmid and Jongman 1984, 97.
81. Kahneman and Tversky 1996, 582.
82. Scheuer 2004.
83. Roman, December 1, 2003.

5

How Terrorism Expedites Organizational Demise

Thus far, I've served up a bounty of multi-method evidence for why rebels should oppose terrorism to achieve their political goals. The political outcome is crucial for assessing the utility of terrorism. After all, terrorism is widely understood from the Oxford English Dictionary to the Federal Bureau of Investigation as a political tactic intended to coerce concessions.[1] Even religiously inspired militant groups issue demands.[2] The Hamas charter asserts, for example, that the Islamist group is dedicated to not only spreading "Islam as the way of life," but also to achieving the "liberation of Palestine."[3] Hezbollah's professed goal is not only to expand Shiite influence in the Levant, but to repel foreign invaders. Salafist terrorists like Lashkar-e-Taiba and Al Shabaab say they want to not only spread their religious ways, but to pressure the governments in Indian Kashmir and Somalia, respectively, to cede political control. But critics might counter that perhaps some militant groups are driven by an alternative incentive structure. I'm sympathetic to this line of argument. In previous work, I've explored how some militant groups seem to place greater value on organizational survival than on redressing their grievances by prioritizing process goals over outcome goals.[4] A reasonable question, then, is whether terrorism is strategic for prolonging organizational longevity. No, it's not.

In this chapter, I'll show that militant leaders should oppose terrorism not least because attacking civilians, especially in large numbers, expedites organizational demise. Terrorism reduces human capital, the backbone of all organizations. Civilian attacks jeopardize the existence of organizations by eroding their supporters including both active and prospective members of the group. Supporters are critical for survival because terrorist groups are made up of members who depend on local and international assistance. The importance of membership size can hardly be overstated. In fact, it's the foremost determinant of militant group survival as with other types of organizations.

Within the field of sociology, organizational ecologists have established that organizations suffer from a "liability of smallness."[5] Across industries, groups are prone to premature death when they have smaller membership sizes.[6] This finding undergirds the adage of organizations possessing "power in numbers." All else being equal, groups gain strength and durability as membership rosters grow.[7] For a variety of reasons discussed below, attacking civilians erodes supporters and thus stymies organizational survival. In its critical pursuit of attracting members and allies, attacking civilians may be the stupidest thing a group can do.

The Immorality of Intentionally Killing Innocent People

The starting point for understanding the negative relationship between organizational survival and civilian attacks is their perceived immorality. For quite some time, killing innocent people has been recognized as bad behavior in international affairs. Founded on concerns about justice, the idea of civilian immunity in war can be traced back to the sixteenth century, when the preeminent theologian of Catholic Europe, Francisco de Vitoria, urged militaries to distinguish between combatants and noncombatants, whom he regarded as the guilty and innocent parties, respectively.[8] A "father of international law," Vitoria reasoned that women and children are illegitimate targets because they abstained from battle.[9] The following century, the Dutch jurist, Hugo Grotius, appealed for immunity to be extended to "men whose way of life is opposed to war-making" such as clergy, agricultural workers, merchants, and artisans.[10] In the seventeenth century, the Swiss philosopher, Emer de Vattel, reiterated that everyone not directly engaged in combat should be spared, including "women, children, feeble old men and the sick" and also "ministers of public worship and men of letters and other persons whose manner of life is wholly apart from the profession of arms."[11] Contemporary philosophers share the moral imperative of civilian immunity. Angelo Corlett stresses, "Targeting of the innocent violates the fundamental intuition that innocent persons ought not be the targets or victims of violent physical attack."[12] John Coady says that the "depth and centrality of the prohibition on intentionally killing the innocent" is a "touchstone of moral and intellectual health" whose rejection would amount to "an upheaval in our moral perspective."[13] Hugo Slim points out that the ideal of civilian immunity has "remained resilient in human consciousness and . . . continues to play on a universal conscience."[14] In addition to philosophers, national leaders also acknowledge that civilians require special status when it comes to target selection. "Murdering the innocent to advance an ideology is wrong every time, everywhere," President George W. Bush opined in his farewell address.[15]

This moral ideal of civilian immunity has been the bedrock of international humanitarian law since the nineteenth century. The 1864 Geneva Convention regulated warfare by establishing protections for wounded combatants.[16] The 1868 St Petersburg Declaration pronounced, "The only legitimate object which States should endeavor to accomplish during war is to weaken the military forces of the enemy."[17] This "principle of distinction" between military and civilian was then encoded in the Hague Conventions of 1899 and 1907.[18] The World Wars in the first half of the twentieth century deviated massively from this targeting ideal, but spurred stronger international laws to safeguard civilians and punish transgressors. Most crimes tried by the International Military Tribunal at Nuremberg were against civilians. The Fourth Geneva Convention of 1949 and its Additional Protocols in 1977 provided unprecedented civilian protections. Article 27 of the Geneva Convention requires troops to refrain from operations that might harm civilians and to take precautionary measures for their safety.[19] The Protocol requires all parties in conflict to "direct their operations only against military objectives," as "the civilian population...shall not be the object of attack."[20] Today, it's widely recognized that all belligerents—governments and non-state actors alike—should adhere to the principle of distinction by avoiding civilian targets to the extent possible.[21] This moral code against intentionally harming the innocent approaches a universal norm; psychologists have found that the vast majority of people around the world feel guilty about harming others when they're seen as having done nothing to deserve the pain.[22]

High Audience Costs of Harming Civilians

Because civilians are seen as sacrosanct, or at least exceptional, attacking them carries what political scientists call "audience costs." What this means is that offenders get punished for unpopular behavior usually in the form of reduced support.[23] To begin testing the audience costs to groups for attacking civilians, I analyze a sample of 118 organizations operating in sixteen Middle East and North Africa countries from 1980 to 2004. The Minorities at Risk Organizational Behavior (MAROB) dataset codes these groups in terms of their tactical choices as well as the governmental responses. Because of the widespread norm against civilian targeting, I examine whether governments are more likely to respond harshly to militant groups when they violate it. To operationalize these simple tests, the independent variable is a dichotomous measure of whether the group committed "attacks on civilians," which the coders define as engaging in "terrorist activities." For the dependent variables, I test the effect of terrorist activities on the likelihood of two kinds of government responses. The first is whether the government used "repression" against the group. And the second is whether the government used "lethal violence" against the group. Table A.7 presents the results. As anticipated, governments are significantly more likely to use repression and violence against a militant group when it engages in terrorist activities by attacking civilians. The magnitude of the effects is noteworthy. Governments are over twice as likely to use repression and over four-times as likely to use lethal violence against a group when it attacks civilians. To probe the valence of these findings, I lagged the independent variable and controlled for all sorts of potential confounds, including with group fixed-effects. Across model specifications, the evidence affirms that governments are apt to use highly punitive countermeasures against groups when they violate civilian immunity. This harsh response is unsurprising from the vantage of social contract theory, which envisions the role of government as protecting the people and their property.

The immorality of intentionally killing innocent people has both direct and indirect costs for membership size. Not only does attacking civilians elevate the risk to current members, but the anticipated cost of participation reduces the incentive for future ones to join. In general, we would expect diminished recruitment whenever the costs of participation grow. As the political scientist, James DeNardo, notes in his book on protest, "The recruiting process is complicated by calculations about the likelihood of repression," so "the movement's support will decline continuously as its tactics become more violent."[24] Of course, the moral repugnance of killing civilians dissuades prospective members in addition to their concerns over physical safety. As the community organizer, Saul Alinsky, eloquently put it, "Men will act when they are convinced that their cause is 100 percent on the side of the angels and that the opposition are 100 percent on the side of the devil."[25]

The April 19, 1995, Oklahoma City bombing illustrates how civilian attacks reduce membership rosters by eliciting government crackdowns, scaring recruits from joining, and turning off participants with a moral conscience. In the aftermath of the largest domestic terrorist attack on U.S. soil, the number of militia groups dropped from a peak of 370 in 1996 to 68 in 1999. As the Southern Poverty Law Center remarked, the Patriot movement became "a shadow of its former self" thanks to the McVeigh attacks. Membership rosters plummeted because the government made hundreds of arrests, prospective recruits feared the same fate, and the moral repugnance of killing civilians wasn't a good look, especially the nineteen children at the Kids Day Care Center in the Alfred P. Murrah Building. A former militia member explained why membership was hemorrhaging after the attack: "Most of the militia people don't view Tim McVeigh as a hero. He's a killer of innocent people. I don't think there's much disagreement on that."[26] Historically, violent excesses are among the main reasons defectors have left White Supremacist and other radical groups. "I felt ashamed after committing violence against people who had not done anything to me," as one former skinhead acknowledged.[27]

In classes on political violence, I sometimes ask my students to consider the following thought experiment to grasp the negative effect of terrorism on participation. Imagine you're a student at a university who's tired of being told that all the tennis courts are taken by the varsity team. To pressure the administration into creating more tennis courts for non-varsity players, you pass around an open letter urging friends to endorse this request. Now imagine going the violent route instead, asking classmates to join you in setting off bombs throughout campus until the administration caves. Under which scenario would you expect to attract more allies? Surely, many more students would sign the cost-free letter than blow up the university because of both their moral inhibitions and fear of getting into massive trouble. This relationship between membership size and anticipated costs isn't just speculative. The political scientists, Erica Chenoweth and Maria Stephan, demonstrate that less violent resistance groups have historically attracted many more members because the "the moral, physical... barriers to participation are much lower."[28] While tiny segments of the population are sometimes tempted to partake in violence against the population, the net effect is to repel participation, reducing membership rosters, and ultimately organizational longevity. When Neo-Nazis stage a rally, counter-protesters typically outnumber them by orders of magnitude. And when Nazis kill people, the backlash is even greater. After James Alex Fields plowed his car in August 2017 into a bunch of pedestrians in Charlottesville, Virginia, counter-protesters outnumbered Alt-Right rallies so badly they cancelled their upcoming events.[29]

Local support is essential for the long-term survival of terrorist groups not only because locals are the most likely to join, but also because they can provide critical assistance. After the war in Bosnia in 1995, for example, the village of Donja Bočinja became a safe-haven for scores of Wahhabi fighters to set up shop under the guise of farmers, earn citizenship, marry local women, befriend politicians, and thereby insulate themselves from U.S.-led countermeasures. The history of terrorism shows that civilian attacks put local support at risk. The Front de libération du Québec (FLQ) "drove Quebecois public opinion away from the group" when it strangled a guy to death in October 1970.[30] When ETA bombed public transportation around Madrid in the summer of 1979, thousands of Basque residents demonstrated against the group.[31] Sympathy for ETA in the Basque Country plunged more than half, falling from 48 percent to 23 percent in response to the carnage.[32] When a bomb intended for the Royal Ulster Constabulary killed eleven bystanders in Enniskillen, Northern Ireland, in November 1987, the IRA's own newspaper said it was a "monumental error" that would "strengthen the IRA's opponents."[33] After a PIRA bomb killed a couple of children in the English town of

Warrington in March 1993, sympathizers in the Republic of Ireland became "more likely to withhold support or even actively oppose the republican movement."[34] The Real Irish Republican Army, a PIRA splinter group opposed to the Good Friday Agreement, committed an even bloodier attack in August 1998, when a car bomb killed twenty-nine people in a crowded shopping area in Omagh, Northern Ireland. The victims included six teenagers, six younger children, a mother pregnant with twins, and tourists from Ireland and Spain. The bloodshed was costly not only to the victims, but also to the group. In fact, this massacre spelled its demise. Community members who shared the same political goals demanded that RIRA disband in what the *BBC* described as "a wave of revulsion triggered by the carnage at Omagh."[35] Such terrorism boomerangs are hardly confined to the West. Just ask Sikh terrorists who killed hundreds of Hindu civilians in the late 1980s, leading to "a noticeable loss of support from Sikh masses" and the elimination of hospitable hideouts.[36] When the Lord's Resistance Army escalated against Ugandans in early 1991, "The LRA's strategy backfired as civilians increasingly turned away from the rebel movement."[37] The so-called Luxor massacre of November 1997 had an equally counterproductive effect on al-Gama'a al-Islamiyya. Six members slaughtered scores of tourists in southern Egypt, resulting in a nationwide backlash that starved the perpetrators of local sympathizers.[38] The Armed Islamic Group (GIA) also self-destructed by killing the masses. Nadia Chaabani, a wife of a GIA leader, vividly captures the group's self-inflicted downfall: "When the killings became commonplace, when heads of neighbors rolled in front of their houses, and the school they'd been so proud of was destroyed, they realized they needed to come to their senses. Once they understood what evil the GIA was capable of, they turned against the terrorists and demanded weapons so they could fight back."[39] She elaborates in her autobiography, "The Islamists had a large segment of the population on their side in the beginning, but because of their ignorance they weren't able to hold on to those sympathizers."[40] According to Al Qaeda strategist, Abu Musab al-Suri, this backlash was "the greatest failing of the entire jihadist experience without exception."[41] Ironically, the same fate befell Al Qaeda in Iraq (AQI), which came to be seen in its country by mid-2005 as *takfir* for blowing up mosques and beheading civilians, mortally wounding AQI's standing even among many local Sunnis. Throughout the world, the history of terrorism abounds with organizations that "imploded" from a loss of local support due to killing civilians, especially in large numbers.[42]

Organizations sacrifice international support as well by attacking civilians. This support is critical for the longevity of terrorist groups in two main ways. First, international support helps groups to continue the resistance with additional manpower, money, weapons, safe-havens, expertise, and alibis. Even so-called domestic terrorist campaigns generally have an international dimension. Most Aum Shinrikyo members came from Russia, not Japan; the Oklahoma City bombers had links with neo-Nazis in Europe; the modal Al Qaeda and ISIS attacker in the West is steeped in online propaganda, de-territorializing the nature of contemporary terrorist support. In a globalized world, external support is so fundamental that some scholars think we should abandon the lexicon of "domestic" terrorists altogether given their international ties.[43] Secondly, international support is essential for groups to successfully transition into legitimate, enduring actors on the world stage. The historical record brims with organizations that have lost and gained international support depending on their restraint towards civilians. The Al Qaeda affiliate in Iraq suffered substantial audience costs in November 2005 when it blew up three hotel lobbies in Amman, Jordan. The political scientist, John Mueller, points out, "The main result . . . was to outrage Jordanians and other Arabs." Polls conducted throughout the Muslim world confirm that the hotel bombings decimated support for Al Qaeda, its affiliates, and their methods against civilians.[44] Similarly, the Beslan school siege in September 2004 that killed 186 children was an organizational disaster for Chechen groups. A spokesman lamented: "A bigger blow could not be dealt upon us . . . People around the world will think that Chechens are monsters if they could attack children."[45] Some rebel leaders have opposed terrorism in order to gain international legitimacy.[46] One study analyzed a sample of 103 rebel groups around the world from 1989 to 2010; those that eschewed terrorism were about 20 percent more likely to successfully lobby international bodies like the UN to gain official legitimacy. In Burma, for example, the Karenni rebels issued a statement highlighting how they incorporated "all international treaties and Geneva Conventions" into its charter. And in Indonesia, the Free Aceh Movement courted international humanitarian bodies: "We believe in and uphold the Law. Under no circumstances does GAM purposely target civilians or their property."[47] There's no shortage of evidence that safeguarding civilians has historically helped groups to grow their membership rosters by attracting more recruits and allies both locally and internationally.

The Implosion of Islamic State

ISIS never got the message. This group incurred substantial audience costs from brutalizing civilians, provoking crippling repression and violence against the group. In the introduction, I noted how ISIS begged the United States and its allies not to intervene against the fledgling caliphate.[48] As I predicted from the moment Baghdadi declared the caliphate at the al-Nuri mosque, the terrorism backfired. The U.S. and France began to pound the daylights out of ISIS after it slaughtered their civilians. The beheading of the American journalist, James Foley, in August 2014 and the *Charlie Hebdo* attack in January 2015 convinced Barack Obama and Francois Hollande to devote their presidencies to crushing the group. This basic pattern of organizational self-destruction repeated itself throughout the world. When ISIS beheaded twenty-one Egyptian Coptic construction workers in Sirte, Libya, the following month, President Abdel Fattah el-Sisi declared in a televised speech it was "time to punish these murderers" before authorizing airstrikes on terrorist targets in Libya.[49] Belgium joined the air campaign in Iraq after an ISIS fighter shot up the Brussels Jewish museum in May 2015.[50] Canada joined the munitions party, dropping 500-pound laser-guided bombs on the group in Fallujah after an ISIS fighter rammed his car into a shopping center in Quebec.[51] Prime Minister David Cameron authorized blistering strikes against ISIS the next month after a soldier of the caliphate went on a shooting rampage in a tourist resort outside of Sousse, Tunisia, killing thirty British vacationers.[52] When ISIS blew up the Russian Metrojet Flight 9268 over the Sinai in October 2015, President Vladimir Putin declared, "We will find them anywhere on the planet and punish them."[53] Putin wasn't kidding: in response to the plane bombing, he expanded his strike force in Syria by thirty-seven planes, so Russian jets could fly sixty to seventy sorties a day, killing hundreds of ISIS members a month.[54] Similarly, the December 2015 attack at a Christmas party in San Bernardino, California convinced Obama of the "great sense of urgency" in hitting ISIS "harder than ever."[55] ISIS attacks turned even erstwhile friends into enemies. For over a year, Turkish president, Recep Erdogan, had defied Obama's appeals to join the anti-ISIS coalition. Prioritizing regime change in Damascus over the fight against ISIS, Erdogan allowed thousands of jihadists to flood into the caliphate. Even Turkey began to resist ISIS, however, when the group launched terrorist attacks inside the country. As payback for the June 2015 bombing in Diyarbakır, Turkey, granted American warplanes airbases to pummel ISIS in what the White House described as a "game-changer."[56] After the Suruç bombing the following month, Turkey for the first time carried out airstrikes against ISIS positions in Syria. As Fawaz Gerges notes, ISIS attacks on Turkey boomeranged: "Recklessly, ISIS transformed a neutral regional power into an enemy and forced Erdogan to finally agree to coordinate operations with the United States against the group."[57] After a suicide bomber struck a Shia mosque in eastern Saudi Arabia in early 2016, the Kingdom offered to send ground forces into Syria though the anti-ISIS coalition demurred.[58] By July 2017, the military coalition had destroyed not only the caliphate, but between 60,000 to 70,000 ISIS fighters in the largest counterterrorism campaign ever assembled.[59]

Throughout this organizational train wreck, the international media fixated on the recruitment effect of the ISIS attacks, while ignoring what I call the attrition effect.[60] News stories detailed every known case of individuals radicalized online by ISIS violence, oblivious to the overall negative effect on the membership size of the embattled group. Even at its height, though, ISIS membership size was tiny compared to less violent organizations like the Muslim Brotherhood. Part of the reason has to do with its extreme ideology. But ISIS also suffered recruitment problems due to the costs of participation. As the group's implosion became obvious to everyone except think tank pundits, fighters scampered away from the group in the thousands. To the naïve, ISIS fighters present themselves as apocalyptic warriors who will fight to the last man. But as the axe fell they fled in droves.[61] *USA Today* noted that "Islamic State defections mount as death toll rises" and the *Independent* reported "Isis increasingly resembles a sinking ship, its militants, like rats, are deserting it."[62] Defectors acknowledge fleeing ISIS because of its unbridled violence.[63] In fact, this was the most common reason for leaving the group in the International Centre for the Study of Radicalisation's sample of fifty-eight ISIS defectors.[64] Defectors linked up with less extreme groups like Hay'at Tahrir al-Sham, Nusra, and other Al Qaeda friends which consistently had more local support because of their relative restraint.[65] This was true from the start. When, in December 2012, the U.S. officially listed Nusra as a Foreign Terrorist Organization, protests erupted throughout Syria's moderate opposition groups, with twenty-nine of them signing a petition condemning the U.S. decision.[66] ISIS never got this kind of love either locally or internationally. And when it came time to negotiate the future of Syria in July 2017, international negotiators in both Geneva and Astana excluded representatives from ISIS because of its extremeness. ISIS is often described as an exceptional group.[67] Like other groups, though, it discovered the hard way that unrestrained violence against civilians evokes crushing blows against the organization, limits recruitment, and minimizes both local and international support essential for long-term success.

False Flag Attacks

Civilian attacks are so counterproductive for the organization that often people conclude its members couldn't have committed them. When civilians are killed, conspiracy theories flourish of a so-called false flag attack, whereby suspicion falls on the government for surreptitiously carrying out the violence in the name of the group in order to weaken it. Conspiracy theorists reason that 9/11, for example, must have been an "inside job" because Al Qaeda couldn't possibly be so stupid as to engage in such self-destructive behavior.[68] Similarly, conspiracy theorists believe that Timothy McVeigh and his accomplice Terry Nichols couldn't have been behind the Oklahoma City bombing because of its crushing effect on the Patriot movement. The veteran police captain, Robert Snow, points out: "Despite the convictions of McVeigh and Nichols, a number of conspiracy theories have sprung up among militia members concerning the Oklahoma City bombing. The most common one is that McVeigh and Nichols are innocent and that the federal government actually detonated the bomb...wanting an incident that would give the government an excuse to clamp down on the militias."[69] Predictably, conspiracy theorists also came out of the woodwork after the FLQ assassinated cabinet minister, Pierre Laporte, when the attack dried up support for the group along with the Quebec independence movement. Interestingly, this conspiracy theory irked the terrorist behind it, Francis Simard, who insisted the incident was perpetrated against the government—not by it. Simard once snapped, "I know what happened, I was there!"[70] In his book on apartheid, Padraig O'Malley writes that ANC attacks on civilians were so counterproductive that many observers attributed them to "the state security apparatus, trying to discredit the ANC."[71] In Algeria, the GIA's indiscriminate violence was so self-defeating for the group that many observers thought the bloodshed was perpetrated by its political enemies. As the investigative journalist, Camille Tawil, observed, "Massacres against civilians were often suspected of having been secretly carried out by Algerian security services."[72] Martin Stone, for instance, speculates in his book on Algeria, "It is also entirely possible that certain elements within the ruling elite ordered some of the murders to focus world attention on the Islamic threat."[73] The political scientist, Mohammed Hafez, has written extensively about this conspiracy theory, "The most common charge held that security forces infiltrated armed groups belonging to the GIA and committed these atrocities to discredit the movement."[74] But the false flag allegation arose because the civilian attacks hurt the GIA—not because of any evidence its opponents did them. As Hafez concludes: "The evidence does not support the claim that security forces were the principal culprits behind the massacres, or even willing conspirators in the barbaric violence against civilians. Instead, the evidence points to the GIA as the principal perpetrator of the massacres."[75] Ever since the September 1999 apartment bombings backfired so disastrously against the Chechens, observers have likewise speculated that the attacks must have been carried out by the Russian government.[76] Predictably, right-wing extremists believe the August 2017 car-ramming attack in Charlottesville, Virginia was a left-wing conspiracy to sandbag the cause, given the fallout.[77] In sum, terrorism has been such a disaster for militant groups that they're frequently seen as hapless victims whether their goals are to coerce concessions or just stay alive. As a rule, militant groups should thus oppose terrorism for both strategic and organizational reasons.

Notes

1. Oxford English Dictionary 1971; Federal Bureau of Investigation 1995; United States Army 1990.
2. Pape 2003.
3. Litvak 1998.
4. Abrahms 2008.
5. Freeman, Carroll, and Hannan 1983.
6. Hutchinson, Hutchinson, and Newcomer 1938; Reynolds and White 1997.
7. DeNardo 1985.
8. Primoratz 2007, 71–2.
9. Woods 2005.
10. Primoratz 2007, 76.
11. Ibid. 77.
12. Corlett 2003, 17.
13. Coady 2008, 297.
14. Slim 2007, 2.
15. "Bush's Farewell Speech," January 16, 2009.
16. Robertson 1999.
17. Lovell and Primoratz 2012, 85.
18. Primoratz 2007, 2.
19. Lovell and Primoratz 2012, 69.
20. Ibid. 22.
21. Best 1997.
22. Baumeister 1999, 206.
23. Schultz 2001.
24. DeNardo 1985, 200, 202.
25. Alinsky 1971, 78.
26. Wolfson, June 20, 2001.
27. "Reflections Of A Former White Supremacist," August 28, 2015.
28. Chenoweth and Stephan 2013, 10.
29. Lanktree, August 22, 2017.
30. Cronin 2011, 108.
31. "Partidos," August 1, 1979.
32. English 2016, 207.
33. Raines, November 15, 1987.
34. Drake 1998, 158.
35. McKinney 2007; Connolly 2007.
36. Sing, June 23, 1988.
37. Stanton 2916, 254.
38. Wright 2006, 258.
39. Gacemi et al. 2006, 75.
40. Ibid. 38–9.
41. McCants 2915, 87.
42. Cronin 2011.

43. Ross 1993, 327; Hoffman 1997.
44. Mueller and Stewart 2015, 124.
45. Pape et al., March 31, 2010.
46. Jo 2017.
47. Stanton 2016, 38, 104, 172.
48. Stern and Berger 2015, 159.
49. "ISIL Video Shows," February 16, 2015.
50. Sciutto et al., September 12, 2014.
51. "Canada Launches," November 3, 2014.
52. "Tunisia Parliament," July 25, 2015.
53. Osborn, November 17, 2015.
54. "Russia Intensifies," August 22, 2017.
55. Harris and Shear, December 6, 2015.
56. Yeginsu and Cooper, July 23, 2015.
57. Gerges 2016, 285–6.
58. The Coalition Demurred for Political Reasons.
59. Brown and Starr July 21, 2017.
60. "Islamic State's Media Violence," August 3, 2015.
61. Poole, July 27, 2017.
62. Brook November 29, 2015; Altindal, August 24, 2017.
63. Walsh, April 12, 2016.
64. Neumann 2015.
65. See, for example, Lister 2016, 279.
66. Gerges 2016, 182.
67. Cronin 2016, 87.
68. Knight 2008.
69. Snow 1999, 104.
70. Thanh Ha, January 15, 2015.
71. O'Malley 2007, 208.
72. Tawil 2011, 132.
73. Stone 1997, 194–5.
74. Hafez 2000, 585.
75. Ibid. 586.
76. Satter, February 2, 2017.
77. Cillizza, August 23, 2017.

6

What Smart Leaders Know

In various ways, all the chapters in this part of the book have shown that terrorism obstructs victory. Terrorizing civilians isn't strategic because it evidently impedes government concessions and organizational support. This insight departs from the conventional wisdom in policy and scholarly circles. As we saw in the introduction, think tank pundits spent years extolling the strategic genius of Islamic State for blowing up random victims in soccer stadiums, markets, shops, restaurants, nightclubs, airplanes, airports, buses, bus stations, trains, train stations, hotels, rock concerts, and countless other venues where ISIS fighters chose to quench their bloodlust. Academics who subscribe to the Strategic Model of Terrorism also overestimate its value, often due to "lumping," that is, failing to disaggregate this anti-civilian tactic from attacks against military and other government targets, which are considerably more effective. A handful of astute scholars have recognized that attacking civilians negatively impacts the group, but ignore the role of its leaders for this poor tactical choice.[1] The dismissal of leaders is a grave mistake in understanding militant group dynamics. Militant groups are failure-prone unless their leaders grasp the perils of indiscriminate violence. Indeed, recognizing this reality is the first step toward becoming a smart, successful militant leader. The final chapter of this section restores agency to leaders by showing that the quality of militant group violence depends critically on their tactical IQ. Knowing the risks of perpetrating terrorism is necessary though hardly sufficient for victory, as we'll see.

Terrorism Is Not a Weapon of the Weak

Most researchers think that groups attack civilians for reasons that have little or nothing to do with their leadership. In the early 1960s, the journalist Brian Crozier coined the adage that "terrorism is a weapon of the weak."[2] This assumption is almost axiomatic in conflict studies.[3] The logic is admittedly alluring: Lacking the capability to battle military forces, groups turn to terrorism to attack softer targets. After all, it's easier to destroy a market than a tank. As a rule, terrorists are unquestionably weaker than governments. But are militant groups that employ terrorism weaker than those that do not? Some of the worst offenders against civilians—Al Qaeda, the Armed Islamic Group, Boko Haram, Islamic State, the Lord's Resistance Army, the PKK, and the Taliban—can hardly be described as weak militant groups. These groups became more ruthless toward civilians after gaining strength over time— that's certainly true for the Armed Islamic Group, Boko Haram, the Lord's Resistance Army, the PKK, and the Taliban.[4] As Walter Laqueur pointed out decades ago, it would be easy to think of many instances past and present when relatively strong groups have targeted civilians.[5] The political scientist, Ken Booth, asserted more recently, "Terror can be part of the menu of choice for the relatively strong."[6] The sociologist Jeff Goodwin likewise remarks, "There does not seem to be a particularly strong empirical relationship between the organizational strength of revolutionary groups and their use (or not) of terrorism."[7] Based on their fieldwork in Sierra Leonne, the political scientists, Macartan Humphreys and Jeremy Weinstein, "find little empirical evidence" that weaker groups gravitated to civilian targets.[8] A study of militant groups from 1998 to 2005 also finds no basis for the longstanding idea that weaker organizations are inclined to soft targets.[9] And a large-n study on post-Cold War civil wars concludes that "the strength of the rebel group has little bearing on the likelihood of restraint" toward civilians.[10]

My first rule for militant leaders to oppose terrorism recognizes that this tactical choice isn't just an inevitable, automatic result of organizational weakness. Otherwise, there would be no point instructing them to oppose civilian attacks. To empirically investigate this counterargument that terrorism is a weapon of the weak, I analyze the statistical relationship between militant group capability and target selection with multiple measures and datasets. With the MAROB dataset of 118 organizations operating in sixteen Middle East and North Africa countries from 1980 to 2004, I find no consistent association between militant group capability and terrorism. It is said that stronger militant groups possess larger membership sizes, control more territory, and have greater popular support. As you can see from Table A.8 in the Appendix, none of these common proxies for organizational capability is associated with a propensity to attack civilians. In fact, groups with more members and territorial control are especially prone to civilian attacks. In the Mideast and North Africa, it would be more accurate to describe terrorism as a weapon of the *week*. Analysis of the 238 militant groups in the Big Allied and Dangerous (BAAD) database at the University of Maryland casts further doubt on the idea that groups reflexively adopt terrorism out of organizational weakness. With BAAD data of militant groups from 1998 to 2005, we again see no statistically significant relationship between either their membership size or territorial control and their proclivity to attack civilians. This dataset also contains a count variable to measure the number of connections to other groups, another standard proxy of militant group capability.[11] As you can see in Table A.9, isolated militant groups with fewer ties to others have also been no more likely to target civilians.

Finally, I analyzed the ITERATE dataset of international hostage crises between 1968 and 2005 to ensure that groups aren't simply escalating to terrorism over time in response to previous negotiation failures. Such evidence might suggest that harming civilians or in this case captives is due to desperation rather than a hidden cause of it. The data, however, offer no empirical support for this seemingly compelling causal story; militant groups are apparently as likely to release the captives unscathed as they are to kill them irrespective of success or failure in the previous standoff with the government; statistically, the frequency of adopting these tactical decisions is identical. Even more tellingly, groups are actually more likely to become nonviolent toward their hostages upon failing with terrorism than they are to escalate with this tactic upon failing with restraint in a previous hostage standoff; the Pearson's product-moment correlation is 0.691 ($p < 0.0001$) and 0.1268 ($p < 0.05$), respectively. In other words, militant groups are more apt to de-escalate from terrorism than to escalate with it in subsequent hostage-takings. Together, these tests put to bed the oft-repeated idea that militant groups inevitably turn to terrorism because they're somehow weaker and more desperate than others.

Statistics aside, a simple thought experiment illustrates the tenuous relationship between capability and terrorism. Imagine you're frustrated that the local middle school doesn't offer your children a healthier lunch. Would you be more likely to commit terrorism if unable to solicit the support of other parents to change the menu? The question is absurd precisely because the presumed relationship between capability and terrorism is at the very least overstated. And normatively, it removes agency from the protesters themselves. Dispensing with this faulty connection is essential for appreciating the role of the perpetrators, especially the leaders who help to call the shots. As the political scientist, C. J. M. Drake, notes, "One of the main determinants of the effectiveness of a leader is his ability to plan operations and to link them to the political objectives of the group."[12] This means knowing which tactics to select.

What Stupid Leaders Don't Know

Militant leaders aren't fighting for a healthier lunch. They may want the government to stop harassing their people or to provide higher wages or to grant self-determination or to adopt Islam as the official religion. At the very least, they want to strengthen their organization to redress some sense of injustice. Whatever their grievance, stupid leaders are unable to recognize that terrorizing civilians is an almost surefire recipe for failure.

A prime example is the GIA, which jihadists uniformly regard as a total bust.[13] In the 1990s, the GIA tried to achieve an Islamic state in Algeria by becoming increasingly nasty toward the population. The political violence was initially restricted to security forces, policemen, and military personnel. In early 1993, the violence expanded to include government officials, especially members of the quasi-parliamentary National Consultative Council and the Transitional Council. Later that year, GIA attacks became noticeably less discriminate, with a rise in violence against opposition groups, foreigners, intellectuals, and ordinary civilians who were often bombed or executed for no apparent reason.[14] By mid-1995, over forty newspaper and broadcasting journalists had been killed.[15] According to jihadist expert, Quintan Wiktorowicz, the portion of lethal GIA-inflicted violence against civilians climbed from 10 percent to 87 percent between 1992 and 1997.[16] The political scientist Mohammed Hafez estimates that between 1995 and 1998 no more than a quarter of GIA attacks were against military or other government targets, as the group preferred decimating markets, cinemas, cafes, restaurants and other public places. At least sixty-seven civilian massacres took place between November 1996 and July 1999.[17]

When civilian abuses are so widespread, it's hard to imagine the policy isn't sanctioned from the top. Indeed, the GIA's diminished targeting selectivity had nothing to do with faceless structural factors like a lack of members or popular support as most scholars would have you believe. No, the carnage was elective, driven by leaders out of pure stupidity. Journalists reported, "The GIA had a problem at the leadership level."[18] In December 1992, the imbecile Abdelhaq Layada issued a fatwa against "ministers, soldiers, and supporters and anyone who works under them, helps them, and all those who accept them or remain silent."[19] Civilian life got even worse with the ascent of Djamel Zitouni to the helm in late 1994, who sanctioned a fatwa against the mothers, wives, and daughters of security forces, resulting in a period of "openly targeting civilians with no apparent link to the conflict."[20] Zitouni boasted that the key to victory was to "not distinguish" between military forces and the wider Algerian society of apostates.[21] Locals understood that "most of the blame...must go to Zitouni himself," as "the most important" factor for the massacre was "his condemnation of whole swathes of Algerian society."[22] Not to be outdone as a strategic dunce, his successor Antar Zouabri in 1996 called for an even grander bloodbath against innocent people, essentially condemning "the entire population of Algeria as infidels."[23] The GIA rank-and-file observed that their leaders seemed "very pleased" whenever "they killed a lot of people."[24] As one member put it, GIA leaders had simply "taken a wrong path and followed the wrong the strategy," which destroyed countless lives, the group, and its cause.[25]

Like the GIA, Islamic State's violent ways have come from idiotic leadership—specifically, a belief at the top that tormenting random people all over the world is the secret to a permanent caliphate. There's a broad consensus among ISIS specialists that Baghdadi and his deputies deliberately implemented the strategic doctrine laid out in Abu Bakr Naji's 2004 internet hit, *Management of Savagery*. Circulated in PDF format online, *Savagery* offered a political roadmap that the ISIS leadership meticulously followed. Malcom Nance notes that ISIS leaders have been "voracious consumers" of this sick manifesto, which is why they "behead innocent people, kill whom they please, and rape and sell women into slavery... it's not pleasure so much as it is written doctrine on how to conduct combat."[26] Jessica Stern and J.M. Berger remark that ISIS was so savage to civilians because the leaders were consciously "following the blueprint in the *Management of Savagery*."[27] Stern and Berger emphasize, "The shocking violence... owed a debt to *Management of Savagery*, the jihadist tract that heavily influenced ISIS' strategy."[28] Shiraz Maher agrees that "*Management of Savagery*... presents the ISIS world-view."[29] So does Hassan Hassan, who notes "*Savagery* is at the core of ISIS ideology."[30] The manifesto was not only widely read among ISIS leaders, but a core text in the curriculum and at the top of its recommended reading list.[31] The *Washington Post* also recommended the text as a must-read for "people looking for insight into the extremist strategy that inflames the fighters of the Islamic State."[32] The *Huffington Post* likewise told readers, "This document is the blueprint of ISIS's leader Abu Bakr Al-Baghdadi's strategy and led to the beheadings, burnings and drownings staged, filmed and edited with ever more sophistication..."[33] But sophisticated isn't how I would describe the ISIS strategy.

In fact, *Management of Savagery* is a guide for how not to manage a militant group. Naji makes clear that his goal is an "Islamic state which has been awaited since the fall of the caliphate."[34] The key to achieving it, he says, is with lots and lots of violence as "failure will be our lot if we are not violent in our jihad and if softness seizes us."[35] To this end, Naji instructs jihadists not to be too picky about which targets to strike and indeed to strike the broadest range possible to demoralize and exhaust the enemy. He recommends for jihadists to "diversify and widen the vexation strikes against the Crusader-Zionist enemy in every place in the Islamic world, and even outside of it if possible, so as to disperse the efforts of the alliance of the enemy and thus drain it to the greatest extent possible."[36] The secret to success, Naji assures, is to "strike with your striking force multiple times and with the maximum power you possess in the most locations."[37] Civilian targets are the best, though, like "tourist resorts in all of the states of the world" and "all of the banks belonging to the Crusaders" and "oil companies" and "apostate authors" and "Arab Christians" and "Westerners" and "hostages," all of whom "should be liquidated."[38] But these are just a few of his suggested targets. He recommends that militant leaders give jihadists the green light for them to slaughter anyone on their wish-list without approval from higher-ups.[39] These gruesome acts should then be broadcast to all corners of the world to scare the international community into compliance. It occurred to Naji that unbridled bloodshed might provoke a crippling response from target countries. But he's confident they'll stop bothering the terrorists as the pain against the populations mounts, resulting in an Islamic state for the ages. Clueless about how deterrence works, he predicted that target countries will "think one thousand times before attacking us" so as not to "pay the price."[40] Of course, it was Baghdadi and his henchmen who paid the price for following this asininity. To be fair to Naji, though, ISIS strategy came from other sources as well. In addition to *Savagery*, Abu Abdullah al-Muhajjer's *Jurisprudence of Jihad* and Abdel-Qader Ibn Abdel-Aziz's *Essentials of Making Ready (for Jihad)* also inform ISIS strategy, or lack thereof. The key to political success in all these manifestos is a torrent of international violence, including against civilians. The path to the golden age of Islam is to be paved with viciousness and gore, so observe no violent limits. Failure will come only from tactical restraint.[41] Clearly, the problem for ISIS wasn't that it lacked a strategy. It's that the strategic blueprint the leaders worshiped stinks. Not all strategies are strategic or at least effective.

Remarkably, ISIS leaders learned nothing from the mistakes of their Iraqi predecessor. The Islamic State was born out of the ashes of Al Qaeda in Iraq, which imploded when its indiscriminate violence turned Shiite and then Sunni Iraqis against the group around 2005 in the so-called Awakening. Bombing schools, hospitals, mosques, shrines, polling stations, markets, and hotels didn't come about from organizational weakness, but the stupidity of its Jordanian architect. Known for his low IQ even in jihadist circles, Abu Musab al-Zarqawi earned the title of "Shaykh of the Slaughterers."[42] Iraqi-based jihadists tell us Zarqawi favored attacking almost anyone "without taking any account of the consequences for society."[43] As Fawaz Gerges notes, "Al-Qaeda lost Iraq and Abu Musab al-Zarqawi, chief of Al Qaeda in Iraq, was the reason."[44] Zarqawi was killed in a June 2006 U.S. airstrike, propelling Abu Ayyub al-Masri and Abu Omar al-Baghdadi to the top slots until they suffered the same fate, leading to the ascension of Abu Bakr al-Baghdadi.[45] Terrorists who operated under all these leaders say Abu Bakr was "the most bloodthirsty of all."[46] Even before gaining notoriety as the leader of ISIS in June 2014, he was hatching terrorist spectaculars against Iraqi civilians, like the April 2011 suicide bombing of the Umm al-Qura Mosque in Baghdad that killed twenty-eight worshippers.[47] The bad apple didn't fall far from the stupid tree.

What Smart Leaders Know

But not all militant leaders are strategic ignoramuses. At the turn of the twentieth century in Ottoman Macedonia, the Greek guerrilla leader, Alexandros Xanthopoulos, concluded that "indiscriminate killing does harm rather than good and makes more enemies."[48] Around the same time, a senior member of Narodnaya Volya also determined, "Those individuals and groups who took no part in the struggle between the [Russian] regime and those fighting for its overthrow would be treated as neutrals, and their life, limb, and property would not be attacked."[49] In the Tan War, Michael Collins entertained the idea of truck bombing non-combatants but decided that such excesses would be self-defeating.[50] The Chinese communist revolutionary Chairman Mao famously shared his wisdom that political success requires winning over—not terrorizing—the population: "Because guerrilla warfare basically derives from the masses and is supported by them, it can neither exist nor flourish if it separates itself from their sympathies and cooperation."[51] The Argentine Marxist revolutionary, Che Guevara, also warned about the political perils of indiscriminate bloodshed: "It is necessary to distinguish clearly between sabotage, a revolutionary and highly effective method of warfare, and terrorism, a measure that is generally ineffective and indiscriminate in its results since it often makes victims of innocent people."[52] The famed Vietnamese General, Vo Nguyen Giap, wrote extensively about the political risks of harming the population. He prided himself on how his fighters "devoted great attention to the strengthening and development of friendship with the people."[53] Unlike Zouabri, Zarqawi, Baghdadi and other dimwits, these revolutionaries understood to varying degrees that the success of violence depends on wielding it selectively.

Historically, many militant leaders have expressed regret when civilians were harmed due to the fallout. A leader of the Mau Mau insurrection in Kenya noted: "The Murang'a people were very brave fighters but they lacked fighting tactics ... they were conducting day battles in the villages which were resulting in the deaths of many women, children and the old persons."[54] When the attack on the Wimpy bar tarnished the ANC, one of its leaders noted that they had to "end this type of operation" lest it sabotage the cause.[55] Michael "Bommi" Baumann, a founder of the German anarchist organization known as Movement 2 June, denounced the 1977 hijacking of a Lufthansa passenger plane as "counterproductive."[56] After the 1987 bombing of a memorial service in Enniskillen that accidentally killed eleven bystanders, the IRA leadership openly acknowledged how the unwanted massacre hurt the cause: "Leaders of the IRA and its political wing, Sinn Fein, recognize that attacks on civilians undermine political support and international sympathy for their campaign to re-unite Britain's six-county province of Northern

Ireland with the rest of Ireland."[57] In his autobiography from inside the Provisional Irish Republican Army, Eamon Collins conceded that whenever members "accidentally killed an innocent man," it put the movement "in a bad light."[58] The Sinn Fein spokesman, Mitchell McLaughlin, put it mildly when he said that "it didn't help" the Republican cause for PIRA members to kill civilians, as the Warrington attack of March 1993 made clear.[59]

Such expressions of regret shouldn't be dismissed as mere propaganda. When Menachem Begin was told about the civilian casualties in the King David Hotel attack, witnesses say "The color drained from the Irgun leader's face as the death toll mounted."[60] Similarly, reports indicate that when ANC leaders found out about the carnage from the Wimpy bar attack, they were visibly "upset by the indiscriminate bombing of civilians."[61] Expressive physical behaviors serve as what Robert Jervis has dubbed "indices" of intentions, as the signals carry inherent credibility due to their uncontrollable nature.[62] Other clues provide additional evidence that some leaders genuinely oppose civilian bloodshed. The psychologist, Margaret Ann Wilson, created an index of "Apparent Intended Lethality" to apply to militant leaders. Contrary to the widespread belief that all leaders want to inflict maximal carnage, some leaders, such as in the IRA and ETA, score low on her index for both authorizing attacks during non-peak hours and issuing warnings to reduce the civilian impact.[63] These behaviors, she suggests, indicate awareness that the leaders genuinely want to limit civilian harm. In his study on IRA and ETA decision-making, the political scientist Joseph M. Brown elaborates that pre-attack warnings are a useful "measure of militants' casualty aversion" that "transforms an otherwise indiscriminate weapon . . . into a sophisticated strategic instrument, capable of inflicting physical and economic harm without egregious harm to life."[64] And sometimes, signs of casualty aversion pop up in internal militant group documents that were never intended for public consumption. For instance, documents captured in 2015 from Naxalite insurgent leaders reveal their dismay over an attack in Chhattisgarh that killed not only the Indian leader, Mahendra Karma, but also a number of bystanders. One of the documents bemoaned, "We could have won sympathy . . . but the unintentional killings of the innocent allowed political parties and the corporate media to label us as terrorists."[65] Unlike the GIA, AQI, and ISIS, the leaders of these groups recognized that civilian attacks are counterproductive.

Smart Leaders Learn the Costs of Killing Civilians

Of course, smart people don't always choose the best approach to achieve their goals. But they learn from their mistakes. Thomas Edison famously tried 2,000 different materials in search of a filament for the light bulb.

When none worked, his assistant grew exasperated: "All our work is in vain. We have learned nothing." But Edison quipped: "We have learned a lot. We know that there are two thousand elements which we cannot use to make a good light bulb."[66] In the same way, smart rebels learn over time the strategic costs of killing civilians. For instance, the GIA's violent excesses spurred its regional commander, Hassan Hattab, to create an offshoot called the Groupe Salafiste pour la Predication et le Combat (GSPC) that differed in a critical way—its restraint towards civilians.[67] The political scientist, Michael Becker, points out that RIRA leaders learned from the disastrous Omagh bombing, which "contributed to an adjustment of the group's tactical choices," as "subsequent attacks were directed at military and government targets in order to minimize civilian casualties." Again, the change in targeting wasn't due to fluctuations in organizational strength, as Becker notes: "RIRA's aversion to especially violent, civilian-directed attacks had little to do with changes in its capability to carry out such attacks...[but] was driven by a desire to retain the approval of the Irish public."[68] Similarly, the PKK used to target Kurdish villages that fielded pro-government militias until they realized that such indiscriminate attacks only strengthened opposition to the Kurdish movement.[69] Pye explains how leaders of the Malayan National Liberation Army (MRLA) concluded in the early 1950s that the indiscriminate violence was driving the masses to cooperate with the government, so "After three years of terrorism... violence was to be used with more discretion... The order for the MRLA [was] to be more selective."[70] In 2002, Palestinian leaders published an "urgent appeal" to stop targeting Israeli civilians inside the so-called Green Line due to the apparent political costs: "We call upon the parties behind military operations targeting civilians in Israel to reconsider their policies... We see that these bombings do not contribute towards achieving our national project that calls for freedom and independence. On the contrary, they strengthen the enemies of peace on the Israeli side and give Israel's aggressive government under Sharon the excuse to continue its harsh war against our people."[71] The Al Aqsa Martyrs' leader, Marwan Barghouti, admonished that Israeli civilians should be off limits because historically such indiscriminate violence has been "detrimental to us."[72] The KKK also began to moderate its tactics upon realizing in the 1970s that its overt demonstrations of racism were only costing the group legitimacy and support.[73] Smart leaders learn not only from the plight of their own group, but also from others. Matthew Levitt notes that Al Qaeda's plight after 9/11 convinced Hezbollah leaders "to roll back its international operations and keep its efforts to strike at Israeli targets as focused and limited as possible." Although it will continue to threaten Israel for the foreseeable future, Hezbollah made the "strategic decision" after 9/11 to invest more heavily in its standing militia than its cadre of international terrorists who have historically exhibited less targeting restraint.[74]

Even within Al Qaeda, some leaders have come to appreciate the stupidity, if not immorality, of indiscriminate violence, though it has been a tragically slow learning curve. As one of the first members on Al Qaeda's top council, Abdel-Qader Ibn Abdel-Aziz played an important role in the "strategy" of Baghdadi and other ISIS leaders to adopt unbridled violence. Over the years, though, the ISIS mentor saw the ignorance of his advice and openly advocated against civilian attacks in both the Muslim world and the West.[75] Osama also exhibited signs of learning. He used to brag in the late 1990s, "We do not differentiate between those dressed in military uniforms and civilians."[76] The Al Qaeda founder emphasized, "He is the enemy of ours whether he fights us directly or merely pays his taxes."[77] But the implosion of his Iraqi affiliate, starting around 2005, revealed to him the costs of indiscriminate violence, especially against Muslim civilians. U.S. Navy Seals discovered a trove of letters from his Abbottabad compound in May 2011, detailing his reservations about the strategic value of such bloodshed. In an October 2005 letter to Zarqawi, Bin Laden's deputy, Ayman al-Zawahiri, told the AQI leader that the violence was backfiring on the cause and organization.[78] The team of West Point researchers who shared this letter with the public noted that Bin Laden and Zawahiri evidently came to appreciate that "indiscriminate attacks" were "a liability, not an asset."[79] Bin Laden remained dissatisfied with Zarqawi's unrestrained violence even after the Zawahiri letter. In December 2005, Atiyah Abd al-Rahman wrote another letter at Bin Laden's behest excoriating Zarqawi for continuing to make "the mistake of lack of precision."[80] Atiyah warned Zarqawi not to pursue the same path as the tactically incontinent Algerian jihadists of the 1990s, who "destroyed themselves with their own hands."[81] Bin Laden and his closest associates now saw with unprecedented clarity "the damage to the Al Qaeda brand that killing civilians had achieved."[82] In September 2013, Zawahiri published "Jihadist Guidelines," where he acknowledged his learning from the past, especially the Iraqi lesson to avoid indiscriminate bloodshed.[83] When it crept into Syria under the guise of the Al Nusra Front and then Jabhat Fateh al-Sham and ultimately Tahrir al-Sham, Al Qaeda would internalize this lesson of restraint towards civilians, helping to outplay ISIS in the long game. In a December 2012 interview in *Time* magazine, the Nusra leader Abu Adnan sought to differentiate the group from the Iraqi affiliate: "America has called us terrorists because it says that some of our tactics bear the fingerprints of Al Qaeda in Iraq, like our explosives and the car bombs. We are not like Al Qaeda in Iraq, we are not of them."[84] For other groups, the learning happened too late, as leaders of the GIA and al-Gama'a al-Islamiya acknowledge.[85] Mohammed Essam Derbala lamented that his fellow jihadists "not only failed to achieve their goals, but, more importantly, it lost them public support... They had forgotten that armed struggle or jihad was never an end in itself."[86] The history of militant groups thus suggests that at least some leaders learn over time to eschew civilian attacks as the costs become apparent.

Robert Pape and other proponents of the Strategic Model maintain that groups turn to terrorism because they have "learned that it works."[87] My analysis suggests the opposite—not only do groups pay a steep price for harming civilians, but the leaders sometimes learn over time the importance of operatives steering clear of them. Let's return briefly to the MAROB dataset to test these competing views on the 118 organizations operating in sixteen Middle East and North Africa countries from 1980 to 2004. If Pape is correct, you would expect the groups on average to increase their attacks against civilians over time, whereas if I'm right you would expect decreased usage of civilian attacks as groups age. To operationalize this simple statistical test, I use logistic regression analysis to determine whether the age of militant groups since their founding is negatively or positively associated with the odds of committing civilian attacks. To help ensure the results aren't driven by potential confounds, I control for numerous factors that could plausibly impact the propensity of groups to strike civilians, like their organizational capability and ideology. Table A.8 presents the results: As they age, militant groups become significantly less likely to attack civilian targets, perhaps due to learning ($p < 0.001$). An illustrative example are Italian groups whose proportion of attacks against military targets grew from 20 percent in 1970 to 75 percent after 1977.[88]

This dynamic can be modeled formally.[89] Consider Table A.10, which shows the payoffs (p) and costs (c) in the short and long term for civilian and military attacks (top). To better illustrate the effects of time discounting, assume the payoffs of both attacks in the short and long term are the same, but the costs differ. The civilian attack may initially seem attractive because the costs c_{cs} are less than for attacking a military target, c_{ms}; however, a civilian attack's long-term costs c_{cl} are greater in terms of angering the target country, eroding support, sullying the reputation of the group, expediting its demise, and reducing the odds of concessions than for a military target, c_{ml}. For concreteness, consider arbitrary utils that follow this ordinal relationship (middle). Without time discounting, the total benefit over the short and long term is 2 for a civilian attack and 1 for a military attack (bottom). Thus, we would expect all groups to target civilians. But with hyperbolic time discounting of the long term, the utility for a civilian attack would be $1+1/t$, and for a military attack it would be $-1+2/t$, where $1/t$ is the weight given to the future.[90] For $1/t < 2$, one prefers the civilian attack, but for $1/t > 2$, one prefers the military attack. Even with shared utility estimates for both costs and benefits, more experienced leaders may therefore learn to prefer military over civilian attacks simply by gaining wisdom and properly gauging the long-term effects.

Of course, the targeting choices of militant groups are hardly a perfect reflection of leadership preferences. Even when a leader knows the strategic costs of civilian attacks, he must prevent his fighters from perpetrating them. That's the next rule for rebels—to build the organization so it limits wayward operatives from defying their targeting preferences by attacking civilians. That's a major challenge because lower-level members are often indifferent to the political agenda of the group, oblivious of how to advance it, and motivated by other goals altogether. Smart leaders know how to craft the organization to restrain even the most useless members from harming civilians, as we'll see in the second part of this book.

Notes

1. Weinstein 2006, 21 acknowledges, "I highlight the importance of structure over agency." See also Kalyvas 2006.
2. Crozier 1960, 191.
3. Crenshaw 1981.
4. See, for example, Hafez 2000; Abrahms and Mierau 2015; Stanton 2016.
5. Laqueur 1977, 12.
6. Booth 2008, 73.
7. Goodwin 2006, 2034.
8. Humphreys and Weinstein 2006, 430.
9. Asal et al. 2009, 273.
10. Stanton 2016, 117.
11. Horowitz and Potter 2014.
12. Drake 1998, 77.
13. McCants 2015, 87.
14. Hafez 2000, 578, 584.
15. Stone 1997, 193–4.
16. Wiktorowicz 2005, 88.
17. Hafez 2000, 584.
18. Tawil 2011, 85.
19. Hafez 2000, 582.
20. Stone 1997, 114.
21. Hafez 2000, 582.
22. Tawail 2011, 129.
23. Ibid. 125; Wiktorowicz 2005, 88.
24. Gacemi et al. 2006, 71.
25. Sifaoui 2004, 53.
26. Nance 2016, 311.
27. Stern and Berger 2015 , 122.
28. Ibid. 2015, 115.
29. Maher, July 12, 2016.

30. Hassan, February 7, 2015.
31. Ibid.
32. Ignatius, September 25, 2014.
33. "France Must Fight Terrorism."
34. Naji 2006, 4.
35. Ignatius, September 25, 2014.
36. Naji 2006, 20.
37. Ibid.
38. Ibid. 20–1, 34.
39. Ibid.
40. Ibid. 31–3.
41. Gerges 2016, 34, 36.
42. Weiss and Hassan 2015, 10.
43. McCants 2015, 35.
44. Gerges 2016, 19.
45. Ibrahim, 4 April 2010.
46. Gerges 2016, 139.
47. Ibid. 146.
48. Livanios 1999, 206.
49. Primoratz 2007, 19.
50. Boot 2013, 258.
51. Tse-Tung 2005, 92.
52. Guevara 2002, 22.
53. Giap 2001, 62–3, 132.
54. Barnett and Njama 1966, 184.
55. O'Malley 2007, 237.
56. Hoffman 1998, 158.
57. Raines, 15 November 1987.
58. Collins and McGovern 1998, 191.
59. Bishop and Mallie 1992, 285.
60. Hoffman 2016, 299.
61. O'Malley 2007, 237.
62. Hall and Milo 2012.
63. Wilson and Lemanski 2013, 7.
64. Brown 2017, 3.
65. Ghatwai 31 May 2015.
66. Furr, 6 June 2009.
67. Tawil 2011, 13.
68. Becker 2017, 225.
69. Marcus 2009.
70. Pye 1956, 105.
71. "Urgent Appeal to Stop Suicide Bombings," June 20, 2002.
72. *Yedioth Ahronot*, September 2, 2001.
73. Chalmers 1981, 405.
74. Levitt 2013, 359.

75. Gerges 2016.
76. "World: America's Clinton Statement in Full," August 26, 1998.
77. Richardson 2007, 6.
78. Bergen 2006, 367.
79. Lahoud 2010.
80. Scheuer 2011, 147.
81. Al-Rahman 2005.
82. Basil and Shoichet, December 22, 2013.
83. Gerges 2016, 243.
84. Abouzeid December 25, 2012.
85. Gerges 2016, 100; Sifaoui 2004, 53.
86. Gerges 2016, 100.
87. Pape 2003, 350.
88. Della Porta 2006, 123.
89. Abrahms, Beauchamp, and Mroszczyk 2017.
90. This discrete model can also be considered a short-hand for a continuous-time model where we integrate the hyperbolically discounted utility from now until the time horizon of the subject. The qualitative results will be the same as exponential discounting.

Rule #2
Restraining to Win

7

Principal–Agent Problems in Militant Groups

The three rules for rebels correspond to each part of this book. The previous part shared the first rule—for the militant leader to understand the costs of terrorizing civilians. This insight is necessary but insufficient for victory because he must still get lower-level members to abide by it. Militants don't magically steer clear of civilians just because the leader snaps his fingers. As we saw in the last chapter, members sometimes attack the population even when their leader is fully aware of the political toll. The second rule for rebels is to restrain operatives from committing terrorism when they may neither recognize the political costs nor even care about them.

Like all organizations, militant groups suffer from what economists call a principal–agent problem, where the leaders can be understood as principals and lower-level members as agents. A principal–agent problem is another way of expressing why the behavior of members often runs counter to the preferences of their leaders and hence the official goals of the group. Principal–agent problems are endemic to organizations because members are often incompetent and pursue their own personal goals upon joining.[1] Consider the owner of a bus company, who wants her drivers to follow the route to maximize profits. This objective isn't always followed because drivers sometimes get lost or favor their own interests on the job, like picking up friends along the way to treat at McDonalds.

In the same way, militant leaders must contend with the fact that many fighters are unable to carry out their wishes or are driven by other ones altogether at odds with the political program. Not only are fighters often political dunces unable to comprehend the fallout from terrorism, but they have stronger personal incentives to commit it. Militant leaders must be adept at preventing the rank-and-file from committing indiscriminate violence because there's an inverse relationship between a member's position within the organizational hierarchy and his incentives for harming civilians.[2] In this part of the book, I'll explain the principal–agent problem facing militant leaders before showing how they can overcome it. The second rule for rebels is grasping this organizational predicament and minimizing it. Smart leaders know not only the political costs of civilian attacks, but how to restrain their members from perpetrating them.

Apolitical Rank-and-File

A basic truism of militant groups is that lower-level members are generally less politically-minded than their leaders. Political indifference—even cluelessness—abounds in the rank-and-file. In his demographic study of IRA members, Robert White found that nearly half of those interviewed were unaware of the discrimination in Northern Ireland against Catholics despite the salience of this issue in official communiqués.[3] At the American detention facility in Iraq known as Camp Bucca, about 80 percent of the jihadists were assessed as "largely ignorant about Islam."[4] The political Islam expert, Olivier Roy, estimates that around 70 percent of jihadists have "scant knowledge of Islam."[5] As the psychologist, Marc Sageman, points out, "People assume that the jihadis are well educated in religion. That is not the case...Their religious understanding is limited."[6] A report on the Liberian civil war in the mid-1990s likewise concluded: "Many fighters seemed to have no idea why they were fighting at all...When asked why they were killing their own people, they would often mumble something about being freedom fighters, but could not explain any further."[7] A study of the Malayan National Liberation Army also found "a lack of political sophistication among the recruits." According to one assessment, "None of them had fully developed political philosophies."[8] A German terrorist in the 1970s acknowledged his own political stupidity, "Most of the comrades of my group called themselves anarcho-trade unionists and so did I, although I did not really understand what it meant."[9] Clearly, members can't be expected to serve the political agenda of the group when they're not even familiar with it.

This doesn't mean the rank-and-file is irrational, so much as driven by an alternative incentive structure. Compared to their leaders, lower-level members are typically more interested in the social benefits of participating in the militant group than in its political platform. Within organization theory, a well-known distinction is made between rational and natural systems models. The former conceives of participation as instrumental to achieve the official goals of the group whereas the latter sees the benefits of participation as an end in itself. Notably, the natural systems model does not assume irrationality; it assumes members derive utility from the participation itself, independent of any other outcome.[10] As I argued in a 2008 paper called "What Terrorists Really Want," the natural system model is a better lens to understand the motives of many militants especially at the bottom of the organization.[11] Some terrorists, like this one, admit it: "Being in the IRA was almost the objective rather than the means."[12] In a sample of 1,100 terrorists in Turkey, those interviewed were ten times more likely to say that they joined the group because "their friends were members" than for its "ideology."[13] West Point researchers from the Combating Terrorism Center found in a sample of 516 Guantanamo Bay detainees that knowing an Al Qaeda member was a significantly better predictor than believing in jihad for turning to terrorism even when a militant definition of jihad was used.[14] A study on Al Qaeda operatives in the West found that they "went to bars and frequented prostitutes more often then they went to the mosque."[15] This apolitical social dynamic is a common feature across militant groups, like the Malayan National Liberation Army, whose recruits joined "as a means for maintaining friendships."[16] Detailed research of German and Italian terrorists in the 1970s has concluded that devotion to other members was the overriding reason for participating.[17] Bommi Baumann, for example, acknowledged: "I had joined the group because of Georg, I knew that he had decided to do certain things and I did not want to abandon him."[18] Case studies on Al Qaeda, Fatah, Hamas, the IRA, Hezbollah, Palestinian Islamic Jihad, Red Army Faction, and the Weather Underground teemed with members claiming to have participated in the armed struggle out of attachment to each other rather than the political agenda.[19]

For this reason, propaganda recruitment often targets the socially isolated rather than those with a demonstrable commitment to the political cause. The terrorism scholar, Ami Pedahzur, remarks that Hezbollah, the PKK, and Chechen and Palestinian groups tend to target young, unemployed men "who have never found their place in the community," not fervent nationalists committed to political change.[20] The political scientist, Peter Merkl, details how Marxist terrorist groups have historically lured unemployed youth with "failed personal lives" who lacked "political direction."[21] The Yemeni specialist, Gregory Johnsen, observes that Al Qaeda in the Arabian Peninsula focuses its recruitment not on committed jihadists, but on "young and largely directionless" socially marginalized Muslim men.[22] Even so-called "lone wolf" terrorists are usually connected to a wider terrorist network. Plots in Europe, Asia, and the U.S. were initially labeled the work of individuals with no operational ties to ISIS until the discovery of direct communications with members of the group.[23]

Years ago, I presented some research to the National Counterterrorism Center on the primacy of social ties among the rank-and-file. A snarky guy in attendance asked me why these low-level terrorists don't just join bowling clubs instead. But bowling clubs wouldn't be nearly as efficient at building affective ties. In a variety of contexts, scholars have found that the danger of committing violent acts strengthens social relationships in groups. In one of the first analyses of gangs, Frederic Thrasher pointed out that the violence serves a critical function in the bonding process, especially among newer members.[24] A landmark study of St. Louis-based gangs in the 1990s highlighted how violence "serves to reinforce solidarity among gang members."[25] In his study of atrocities, Mark Osiel describes how mass murder, as in World War II, has the perverse effect of "securing a spectacular measure of . . . cohesion."[26] Peggy Reeves Sanday's study of gang rape at college campuses stresses how the violent act helped the men to bond.[27] Wartime sexual violence also has a centripetal effect on the perpetrators, as the political scientist, Dara Kay Cohen, documents: "Rape may be used by combatants to increase intragroup social ties where they are lacking . . . violence can create cohesion among combatants."[28] A similar dynamic infuses social movements. The sociologist, Sidney Tarrow, noted of nineteenth-century French labor protesters, "As they faced off against hostile troops or national guardsmen, the defenders of a barricade came to know each other as comrades . . . and formed social networks."[29] The sociologist, Doug McAdam, mentions how the social bonds of civil rights activists strengthened in the Freedom Summer of 1964 as they adopted riskier forms of participation.[30] The community organizer, Saul Alinsky, says protesters often escalate tactically to build group cohesion.[31] Donatella Della Porta found in her demographic study of Italian and German terrorists that "shared involvement in risky activities strengthened the loyalty ties inside the peer groups of comrade-friends." Left-wing militants attest that "the danger we lived together, as soon as it was over, brought about moments of great closeness," so "anybody who fights is one of us."[32] The post-colonial revolutionary, Frantz Fanon, was apparently on to something when he prescribed armed resistance to foster communal solidarity against the oppressors.[33]

For many militants, the risks of fighting aren't a means for solidarity but an end in itself. Many IRA members describe the excitement of participation over the putative goal of Irish independence.[34] As one member put it, "For a lot of us, it was a big adventure."[35] The Malayan Communist Party attracted recruits by "providing excitement."[36] Terrorists have historically flocked to the action regardless of their political background. In the 1970s, thousands of terrorists from dozens of countries and organizations descended on training camps run by the Palestine Liberation Organization. In the 1980s and mid-1990s, the locus of terrorist activity shifted to Afghanistan to train with the Afghan mujahedeen and then Al Qaeda. First-hand accounts from these camps confirm that the terrorists often had little idea or preference where they would end up fighting after the training.[37] Based on her interviews with some of these jihadists, Jessica Stern has likened their adventures to an "Outward Bound" experience for young men seeking excitement.[38] The Saudi religious scholar, Musa al-Qarni, confirms that many of the Arab youth who went to fight in Afghanistan "had no knowledge of prayer or the ritual ablution" as "they came only to perform jihad."[39] A study of foreign fighters in Syria finds that they tend to be "thrill seekers," looking to take "a harrowing adventure" and "engage in action while enjoying a certain level of impunity."[40] Research by David Malet suggests foreign fighters throughout the world have historically been even more adventure-driven than local fighters or what Brian Jenkins calls "action-prone."[41] Unsurprisingly, then, foreign fighters have tended to inflict disproportionate bloodshed. After Saddam Hussein was toppled, foreign fighters in Iraq made up less than 10 percent of the insurgents but were responsible for over 90 percent of suicide bombings, which are extra lethal.[42] In his examination of more than 500 suicide bombings during the occupation of Iraq between March 2003 and August 2006, Mohammed Hafez found that "many, if not most" of the perpetrators were foreigners with a significant number from outside the Arabian Peninsula.[43] More recently, jihad expert, Thomas Hegghammer, points out that "foreign fighters are overrepresented, it seems, among the perpetrators of the Islamic State's worst acts."[44] ISIS propaganda has catered to the apolitical jihadist. A content analysis concluded: "Come to the Islamic State, is the message. There is fun here!"[45] David Rapoport discovered that terrorist groups historically die out within a generation, perhaps as members age and the allure of adventure fades.[46]

Even when leaders understand the political risks of terrorism, the rank-and-file may not care about them. Like all organizations, militant groups are internally heterogeneous social units. In general, the leaders are more politically motivated than the rank-and-file, which often prioritizes the benefits of participation itself as anticipated in the natural systems model. This divergence in priorities creates an obvious principal–agent problem for leaders intent on wielding violence in a manner that optimizes their political objectives.

The Incompetent Rank-and-File

Even when lower-level members are committed to the political platform, they're still usually more ignorant than their leaders about how to achieve it. Militant groups are invariably bedeviled by an educational discrepancy between leaders and foot-soldiers. For starters, the leaders tend to be the oldest members of the group, so they're the most likely to have observed the strategic fallout of attacking civilians. By contrast, low-level members are typically the youngest members with the least exposure to the actual effects on the organization of the attacks. In his study of Egypt-based militant groups in the 1970s, the sociologist Saad Eddin Ibrahim finds that the leaders were between fourteen and sixteen years older than their fighters, depending on the organization.[47] The FBI agent Thomas Strentz determined that American domestic terrorist leaders in the 1970s were about fifteen years older on average than operatives.[48] Marc Sageman concludes that Al Qaeda leaders have been at least a decade or two older than recruits.[49] Latin American leaders have also been noticeably older. Brazil's Carlos Marighella was fifty-eight at the time of his violent death. Abimael Guzman, who led the Shining Path of Peru, was the same age at the time of his arrest. The leader of the Uruguayan Tupamaros, Raul Sendric Antonaccio, was forty-two when he founded his movement.[50] With age comes experience about which asymmetric tactics work, as this ANC leader acknowledged: "With the exception of a few founding members who had some military experience as a result of their being veterans…most of the regional commands consisted of military illiterates."[51] Huks leaders in the Philippines attest that combat experience honed their choice of tactics: "Gradually, from engagements with the enemy, we learned guerrilla tactics and became an effective armed force."[52] Similarly, Gene Sharp notes that the leaders of nonviolent resistance campaigns are typically "the most experienced person[s] with deepest insight into the technique, the social and political situation."[53] American civil rights leaders say they learned "often by trial and error" from failure in Albany to success in Birmingham and Selma.[54]

We see a similar age distribution in gangs, as the criminologist Malcomb Klein describes: "In general, the original gangsters are the oldest and these veterans make up the leadership. They are sometimes called seniors. Juniors constituted a middle-aged group, and midgets or babies or some other name would be at the bottom. And within these levels there were leaders of those subgroups."[55] Another major study of gangs reaches the same conclusion: "Age (in combination with length of time in the gang) was…noted as a major criterion that set leaders apart from other members. In particular, the old gangster, or OG, was a role identified with leadership."[56] As you might expect, the oldest members are also the leaders in female-run gangs.[57] Research finds that common criminals tend to become more competent over time, so it makes sense for elders to assume leadership roles.[58] As in gangs, militant group leaders have more wisdom from direct experience than the rank-and-file who are comparatively unseasoned and know little about how to advance the organizational mission, when they even care about it.

The rank-and-file also has less formal education, so they are more ignorant about the experiences of other asymmetric conflicts. Studies on social movements find that leaders are seldom representative of the group, as they tend to be endowed with superior educational capital.[59] Vladimir Lenin, Mahatma Gandhi, Martin Luther King Jr., and Betty Friedman presided over various types of social movements, but each was relatively educated.[60] The sociologist, Ekkart Zimmermann, observes that revolutionary leaders have been "better educated" than the rank-and-file.[61] The political scientist Thomas Greene notes, "Compared with their followers, leftist revolutionary leaders are likely to be better educated, more widely travelled, and drawn from higher social classes."[62] The leaders of Italy's Red Brigades and the Shining Path of Peru were college professors.[63] Even in ISIS, the leaders stood out for their higher formal education. As Will McCants notes about Baghdadi and his advisors: "Although the Islamic State's soldiers might not know Islamic scripture very well, some of its leaders do. The caliph has a Ph.D. in the study of the Qur'an, and his top scholars are conversant in the *ahadith* and the ways medieval scholars interpreted it. There are many stupid thugs in the Islamic State, but these guys are not among them."[64] Of course, there's a major difference between smarts and education. I wouldn't go so far as to say that ISIS leaders are even minimally smart despite their education. Tellingly, Baghdadi enrolled in the College of Islamic Sciences at the University of Baghdad after being denied entry to study law for poor test scores.[65] In *The Islamic State: Disentangling Myth from Reality*, Abu Abdullah Mohamed al-Mansour al-Issawi—the leader of Jayish al-Mujahideen—describes Baghdadi as a mental "mediocrity" with "limited intelligence."[66] For other leaders, though, their erudition extends to the study of political violence. As Pierre Vallières of the FLQ acknowledged: "I studied in particular the writings and deeds of the revolutionaries of our time: Lenin, Rosa Luxemburg, Mao Tse-tung, Castro, and Che Guevara."[67] The ANC leader, Mac Maharaj, self-consciously "took a leaf out of the copybook of the Vietnamese struggle."[68] Mandela studied the FLN for strategy because "the situation in Algeria was the closest model to our own in that the rebels faced a large white settler community that ruled the indigenous majority."[69] Luis Taruc, the commander-in-chief of the Huks, styled his campaign in the Philippines on the Chinese Communist guerrillas, using Edgar Snow's *Red Star Over China* as a textbook.[70] In "Problems and Prospects of Revolution in Nepal," the founder of the Communist Party of Nepal, Pushpa Kamal Dahalm, explicitly draws upon the writings of Chairman Mao and other revolutionaries.[71] A leader named Monimambu of the People's Movement for the Liberation of Angola said: "Our struggle is not an isolated one. We [are] profiting from the experience of others. We must learn from the Chinese, etc. But now the most advanced form of guerrilla struggle is in Vietnam."[72] In his autobiography, Yasser Arafat explains how the Palestine Liberation Organization tried to emulate the post-colonial Algerian revolutionaries.[73] PLO leaders also reportedly studied the tactics of the IRA and Irgun in designing their campaign against Israel.[74] PIRA leaders studied the works of Tom Barry and Robert Taber on guerrilla warfare theory. And leaders in the Red Brigade treated Marighela's "Handbook of Urban Guerrilla Warfare" as "a blueprint" for their own revolutionary strategy.[75] Of course, neither experience nor schooling guarantees success in any career. As we saw in Chapter 3, many militant leaders have drawn incorrect tactical lessons by misreading the history of asymmetric conflict. And Chapter 6 showed how Baghdadi and other ISIS leaders learned to be political losers by following a terrorist manifesto for dummies. The point, rather, is that low-level members tend to be relatively uneducated in terms of both combat experience and schooling. So, even smart leaders must struggle to get them to make the right tactical choice given their relative ignorance about political violence.

Lower-level members are not only less educated than their leaders, but often flat-out stupider. There's no shortage of stupid terrorists, especially at the bottom. Take Najibullah Zazi, who went to Afghanistan to link up with the Taliban. Upon returning to the U.S., he couldn't remember how to build a bomb and was busted for e-mailing his questions to the FBI. "Zazi was a stupid kid, believe me," his uncle recalled.[76] And there's the story of the neo-Nazi in Norway who, in his rush to bomb a synagogue, took a tram going the wrong way and blew up a mosque instead.[77] Pye finds in the Malayan National Liberation Army "a great difference in respect to ability and skill" between leaders and foot-soldiers who had only "limited" aptitude for "cause-and-effect relationships."[78] Ted Robert Gurr also notes that leaders are better at calculating "the utilities of political violence."[79] Leaders often rise to the upper echelon or at least maintain their senior position through a meritocratic vetting process. As one study on protest groups determined, "Individuals often land leadership positions because they are best suited to design and preside over social movement tasks."[80] A member of the Mau Mau insurrection in Kenya describes in his autobiography how "leaders were selected on the basis of their demonstrated abilities."[81] A former PIRA member recounted how a Derry Provisional ascended the organizational hierarchy in the 1970s: "He was articulate and could explain publicly where we were at. That was unusual in the IRA and it secured his place very early on and kept him there because he is still able to think."[82] Members of the Red Army Faction said that Gudrun Ensslin assumed a leadership role alongside her boyfriend, Andreas Baader, because of her obvious intellect.[83] Militant leaders thus suffer from a principal–agent problem not only because subordinates are indifferent to the group's political goals, but less knowledgeable about how to achieve them due to inferior experience, education, and aptitude.

The Perverse Incentives of the Rank-and-File

Another challenge for the leader is that his subordinates have stronger incentives to attack civilians. Revenge is an emotional reaction to a perceived loss that goes beyond any instrumental use other than the sense of righting injustice.[84] Attacking civilians is almost always counterproductive for the political success of groups, but can offer utility in settling a score. Compared to leaders, subordinates are far more likely to have suffered at the hands of the enemy. Militant leaders and their families seldom reside in the most dangerous areas of the conflict zone. A study on protest from the early 1970s remarked: "The leadership is in most situations kept back, out of danger. Indeed, the movement may depend upon the safety of the top leadership."[85] Walter Laqueur made a similar observation around the same time: "The central command of the terrorist movement has sometimes been located abroad; Switzerland, the United States, Lebanon have been centers for movements operating elsewhere. The advantages are obvious: The militant leaders can move about freely without fear of arrest."[86] Just ask Hamas and Taliban leaders in their fancy Doha-based hotels.[87] Osama Bin Laden was watching TV in his upper-class Abbottabad living quarters for five years before his capture, while his fighters sweated it out in Yemen, Somalia, Iraq, and other conflict zones. Tucked away in fortress Alamut in eleventh-century Persia, Hasan-i Sabbah dispatched "Assassins" armed with daggers to stab Muslim notables who opposed him.[88] Because militant leaders seldom deploy to the front-lines their operatives are prone to greater personal traumas and incentives to right them.

Historical accounts confirm that revenge is a powerful motive in militant groups especially among the rank-and-file. The terrorism scholar, Assaf Moghadam, finds that lower-level members of Palestinian groups have been disproportionately motivated by the vengeance impulse.[89] In her El Salvador fieldwork, the political scientist, Elisabeth Wood, discovered that the *campesinos* fought for the utility of vengeance, not political reform, against those who had wronged their family and friends.[90] The Confederate Missouri guerrilla, Bill Anderson, wrote to citizens of Lexington in 1864 of his misdeeds in taking revenge on the Yankees for killing his father and sister: "I have fully gutted [sic] my vengeance... I have tried to war with Federals honorably, but for retaliation I have done things, and am fearful will have to do that which I would shrink from if possible to avoid."[91] A Polish underground fighter during World War II conceded, "Very often these people use the weapons allocated them by the movement without authorization and for personal purposes: to exact vengeance."[92] Another member acknowledged that "life had ceased to mean anything to him after he had lost his sister and his brothers, his parents, his wife and his three daughters in a [Ukrainian] raid [and so he] continued to exist only for the purpose of killing and torturing Ukrainians."[93] An autobiography of a Huks commander underscores the perverse incentives of his fighters to strike civilians: "I admitted that errors were made and that innocent people died... acts of revenge were perpetrated by fighters whose families had been murdered or who had suffered at the hands of collaborators."[94] He elaborates, "Some of the Huks, blinded by hatred and incensed by injustice, committed cruel retaliatory atrocities."[95] After his mother was killed by government troops, for instance, a young fighter reportedly raped and killed a female teacher in the Pampanga Province in a cathartic "counter-atrocity."[96] A Mau Mau guerrilla noted among the rank-and-file a widespread "desire to fight back or retaliate."[97] ANC president Oliver Tambo bemoaned the difficulties of restraining "the relatively untrained and embittered township youths...from attacking soft white targets."[98] A Lebanese woman in the civil war captures the desire for vengeance in conflict zones: "It concerns me when three hundred and sixty-five Lebanese Muslims are murdered. I feel the seeds of hatred and the desire for revenge taking root in my very depths. At this moment I want the [militia] or anybody else to give the Phalangists back twice as good as we got. I would like them to go into offices and kill the first seven hundred and thirty defenseless Christians they can lay their hands on."[99] Similarly, a leftwing Italian terrorist in the 1970s shared his justifications for lashing out at civilians: "If you want to hurt the Christian Democracy in a neighborhood, you need a target...Then you have singled out your victim...he is the one to be blamed for everything...you punish him not only for what he has done but also for the rest. Then you don't care anymore which responsibilities that person has; you ascribe everything to him."[100] "He pissed me off" was the reason offered by a Liberian fighter for shooting someone in the civil war.[101] Killing him didn't advance the cause, but probably made the fighter feel better if only momentarily.

To assess the relationship between anger and terrorism more scientifically, my Northeastern colleagues and I conducted a content analysis with IntelCenter's militant group propaganda video database.[102] IntelCenter is a private contractor based in Alexandria, Virginia, that provides access to thousands of propaganda videos from militant groups around the world. This resource has been used primarily by counterterrorism practitioners, but also some academics.[103] Because civilian attacks aren't instrumental, we hypothesized that they're more likely to be featured in propaganda videos when the spokesman appears angry. To facilitate the investigation, we restricted our sample to the ten most active militant groups in the decade after the September 11, 2001 attacks that had video representation in the IntelCenter database, thereby allowing us to focus on groups that commit many attacks and produce large quantities of propaganda for analysis. This selection criteria narrowed the sample to the following groups: the Taliban, FARC, Tehrik-i-Taliban Pakistan, Boko Haram, Al Shabaab, Al Qaeda in the Arabian Peninsula, Chechen rebels, the Salafist Group for Preaching and Combat, Al Qaeda in the Lands of the Islamic Maghreb, and Al Qaeda in Iraq.[104] Graduate students fluent in both Arabic and English coded the 473 videos to assess whether they're more likely to issue threats against civilian targets than military ones when the spokesman exhibits cues associated with higher levels of anger, like yelling or fist-pounding. Table A.11 shows that even after controlling for numerous potential confounds like production quality and strategic context, militants who appear upset are significantly more likely to threaten civilians.[105]

This analysis sheds light on why lower-level members of militant groups are often more disposed than their leaders to harm civilians. Admittedly, the empirical relationship between emotionality and civilian attacks isn't confined to the rank-and-file. Even leaders get upset and bay for blood. As we saw in Chapter 1, Bin Laden wanted Al Qaeda to attack Americans as punishment for U.S.–Mideast policies. Vicarious pain is real, which is why Islamist groups around the world feature propaganda of heart-wrenching scenes from Gaza. But the vengeance impulse is more likely to be pronounced among those who actually suffer, like lower-level members who don't have the luxury of living away from the battle to make dispassionate tactical calculations. For the rank-and-file, terrorist attacks are often a means of quenching their desire to right a perceived loss even if they end up compromising the political objectives of the group. Militant leaders face a challenge in restraining lower-level members from committing terrorism given their penchant for vengeance notwithstanding the political costs.

Unlike the leaders, low-level militants can also boost their organizational standing by striking civilians. Indiscriminate violence can help the rank-and-file to ascend by establishing a reputation for risk-acceptance. Ahmed Chaabani climbed the ranks of the GIA because he was "so aggressive" and acted as "a daredevil."[106] Based on his interviews with IRA members, the terrorism researcher, Gary Ackerman, found that "the goal of attacks was more often than not about personal glorification and the desire to be seen as a hero amongst one's peers."[107] Upon joining Nusra, new recruits headed straight to the front line to demonstrate to their peers "bravery, dedication, and loyalty to the organization."[108] Mohammed Emwazi, a.k.a. Jihadi John, climbed the ranks of foreign fighters in Syria by virtue of his videotaped beheadings of James Foley, Steven Sotloff, David Haines, Alan Henning, and Peter Kassig. As I've mentioned, these gruesome acts were deeply counterproductive for ISIS. From a personal perspective, though, they propelled Jihadi John up the ladder of ISIS foreign fighters until he got droned in November 2015. Performative violence isn't limited to the rank-and-file of militant groups. Dara Kay Cohen finds that wartime gang-rape can improve the standing of low-level militants by demonstrating their "willingness to take risks."[109] Research on gangs finds that new members participate in gang-rapes not only to bond with other horrible members, but to gain stature.[110] Newest members are the most likely to commit violence in order to show they're "tough enough."[111] They report that shooting someone "gave them rank, higher status, and responsibility in the gang."[112] Criminologists emphasize that a key function of the gang leader is to "set violent limits" in order to "keep younger members from feeling too powerful."[113] This is exactly what smart militant leaders do with their members as well.

111

In this chapter, we saw how militant leaders face a principal–agent problem because their subordinates are often indifferent to the political plight of the group, clueless about how to advance it, and rewarded for acting against its long-term interests by attacking civilians. Clearly, the first rule for rebels—of opposing terrorism—is inadequate if they're unable to restrain their members from harming civilians. In the remainder of this section, I'll explain how rebels build the organization to minimize the principal–agent problem, get fighters to follow their targeting preferences, and thereby boost the odds of victory.

Notes

1. Hawkins et al. 2006; Gould 2006.
2. Abrahms and Potter 2015.
3. White 1992, 83.
4. Stern and Berger 2015, 35.
5. Lerner, August 20, 2017.
6. Sageman 2008, 51.
7. Ellis 2006, 127.
8. Pye 1956, 40, 161.
9. Della Porta 2006, 150.
10. Scott 1981.
11. Abrahms 2008.
12. O'Rawe 2016, 6.
13. Schmid 2007, 12.
14. Felter and Brachman 2007, 24–5, 34.
15. Sifaoui 2004, 65.
16. Pye 1956, 237.
17. Della Porta 2006, 168; Wasmund 1986, 204.
18. Baumann 1979, 185.
19. See, for example, White 1992; Post, Sprinzak, and Denny 2003; Braungart and Braungart 1992.
20. Pedahzur 2005, 137–8, 168.
21. Merkl 1986, 42.
22. Johnsen 2007.
23. Walsh 4 February 2017; Callimachi May 12, 2015.
24. Thrasher 2013.
25. Decker and Van Winkle 1996, 17.
26. Osiel 1999, 155.
27. Sanday 1990.
28. Cohen 2016, 13.
29. Tarrow 1994, 44.
30. McAdam 1990, 70–5.

31. Lipsky 1969, 2–3.
32. Della Porta 2006, 154, 177.
33. Fanon and Philcox 2004.
34. Clarke and Johnston 2001, 47. Toolis 2015, 87.
35. Moloney 2010, 48.
36. Pye 1956, 236.
37. Nasiri 2008, 151, 178, 217.
38. Stern 2003, 5.
39. Lahoud 2010, 19.
40. "Understanding the Jihadists," 19.
41. Malet 2013, 2; Jenkins 1982, 15.
42. Quinn and Shrader 2005; Pape 2005.
43. Hafez 2007, 89, 219.
44. Holland, October 10, 2014.
45. Tarabay et al., March 20, 2015.
46. Rapoport 2002.
47. Ibrahim 1980.
48. Strentz 1988.
49. Sageman 2004.
50. Richardson 2007, 46.
51. O'Malley 2007, 98.
52. Taruc 1967, 22.
53. Sharp and Finkelstein 1973, 465.
54. Barker et al. 2001, 6.
55. Klein 1997, 58.
56. Decker and Van Winkle 1996 , 97.
57. Ibid.
58. Canter 1994, 214, 297–8.
59. Morris and Staggenborg 2004, 175.
60. Ibid. 174.
61. Zimmermann 2012, 312.
62. Greene 1990, 57.
63. Richardson 2007, 46.
64. McCants, August 19, 2015.
65. Morris, February 19, 2015.
66. Gerges 2016, 135.
67. Vallieres 1971, 214.
68. O'Malley 2007, 232.
69. Mandela 2013, 355.
70. Taruc 1967, 33.
71. Weinstein 2006, 32.
72. Barnett et al. 1972, 15.
73. Hart 1989, 113.
74. Enders and Sandler 2012, 18.
75. Drake 1998, 76.

76. Mueller and Stewart 2016, 42.
77. Horgan 2009, 2.
78. Pye 1956, 116, 127.
79. Lichbach 1998, 173.
80. Morris and Staggenborg 2004, 175.
81. Barnett and Njama, 1966, 154.
82. Toolis 2015, 308.
83. Aust 1987, 92–3.
84. Baumeister 1999, 26–7, 132.
85. Sharp and Finkelstein 1973, 637.
86. Laqueur 2016, 85.
87. Havard and Schanzer, May 11, 2017.
88. Boot 2013, 206–7.
89. Moghadam 2003.
90. Wood 2001.
91. Fellman 1989, 60.
92. Zawodny 1978, 282.
93. Ibid. 70.
94. Taruc 1967, 23.
95. Ibid. 159.
96. Ibid. 30.
97. Barnett and Njama 1966, 149.
98. Holland 1990, 29.
99. Tabbara 1979, 54.
100. Della Porta 2006, 174.
101. Ellis 2006, 127.
102. Abrahms, Beauchamp, and Mroszczyk 2017.
103. See, for example, Salem et al. 2008.
104. The sample doesn't include Islamic State because the temporal window stopped in 2012 due to data limitations; its forerunner, AQI, is included.
105. As you can see in the table, videos in which the spokesman appears angry are inversely related to those featuring selective violence threatening military and other government targets as opposed to indiscriminate violence against civilians.
106. Gacemi 2006, 48.
107. E-mail correspondence with Gary Ackerman on February 26, 2017.
108. Benotman and Blake, n.d.
109. Cohen 2016, 3.
110. Amir 1971.
111. Decker and van Winkle 1996, 71.
112. Ibid. 71.
113. Bourgois 2003, 121.

8

Cultivating Task Cohesion

The smart rebel knows a nasty principal–agent problem stands in the way of victory. He has learned the political risks of terrorism, but this knowledge is worthless unless subordinates carry out his targeting wishes. How can he get them to eschew terrorism when they're indifferent to the cause, oblivious about how to advance it, and may benefit personally from subverting it with civilian attacks? The key for militant leaders is to cultivate in the ranks what's known in management as task cohesion.

Task cohesion is the degree to which group members act together to accomplish a common goal, for example, to win a game in sports.[1] A group with high task cohesion is composed of members who subsume their private interests by engaging in collective efforts on behalf of the group.[2] Ted Robert Gurr describes task cohesion as "the extent of goal consensus and cooperative interaction among members."[3] Over five decades of research in group dynamics, organizational behavior, social psychology, and military sociology suggest that organizational performance hinges on commitment to tasks in pursuit of the common goal.[4] This chapter offers what might seem like obvious advice for rebels to cultivate task cohesion in the ranks. But this initial step is crucial and inconsistently applied by leaders to restrain members from jeopardizing the cause with terrorism. Task cohesion in militant groups begins with educating the rank-and-file about their political ends and the optimal means to achieve them. Doing so goes a long way toward limiting the counterproductive attacks that sink so many armed struggles.

Teaching the Ends

In the early 1970s, the sociologist, William Gamson, analyzed scores of American volunteer groups operational between 1800 and 1945 to unearth the determinants of success. An unexpected property was closely associated with victory—the provision of a written document like a constitution or charter articulating the cause.[5] What matters for success isn't the paper on which it's written, but that leaders educate subordinates about the cause so they're able to serve it. An ANC member recalls: "All comrades took courses taught by those with more education ... Besides our regular courses, we met regularly for political discussion."[6] Hezbollah also "steeps its membership" in political education.[7] Historically, the leaders have invested about a year teaching each recruit the ideological goals of Hezbollah. The purpose, writes Ahmad Nizar Hamzeh, is to ensure all members give "support to Hezbollah's cause."[8] In the Ugandan Bush war, National Resistance Army leaders believed "political education was important," so they taught the rank-and-file about the importance of over-throwing the corrupt Milton Obote government.[9] Smart leaders impress on the rank-and-file that they should use force instrumentally in the service of their political objectives, as the nineteenth-century Prussian general, Carl Von Clausewitz, famously prescribed.[10] A leader in the People's Move-ment for the Liberation of Angola put it this way: "Our principle is to combine the military and the political. Everyone must be both political and military together. We know our basic problem is a political one, but it cannot be solved without violence ... There is no difference between political and mili-tary leaders inside now. Every person holding a leadership position partici-pates in both the military and political aspects of the struggle."[11] The Cold War historian, John Gaddis, reminds us that strategy is the alignment of means and ends, so the rank-and-file cannot be expected to select the right tactics without grasping the political purpose.[12]

And yet a surprisingly large number of militant leaders neglect to teach their members why they're fighting in the first place. Accounts of the Mau Mau insurgency in Kenya indicate there was "little or no political education for the militants or the masses," which "resulted in the absence of an overall politico-military strategy," spurring "opportunism, adventurism and defeatism."[13] Field research on Renamo leaders in Mozambique reveals they "did not use political education as a tool to set in place shared beliefs about the purpose of the war and the way in which it should be conducted."[14] Former combatants recall a "dearth of political education," as their leaders "said little about the purposes of the war."[15] As you might expect, militants in Renamo were far more likely to commit indiscriminate violence than those in other East African groups like the National Resistance Army that educated members about the cause. In Uganda, Joseph Kony helped found the Lord's Resistance Army to overthrow president Yoweri Museveni. As leader, though, Kony "never articu-lated these political objectives, leading to debate over whether the LRA had a political agenda at all."[16] The founder of the Aum Shinrikyo doomsday ter-rorist group likewise failed to articulate a political vision beyond paranoid alarmism about the U.S. launching an apocalyptic Armageddon gas attack against Japan.[17] Leaders of the Weather Underground also failed to explain their political point, beyond destroying capitalism. In her autobiography, Susan Stern admits: "The greatest single failure of Weatherman... [was that] we simply represented no alternative to anyone. Once we tore down capitalism, who would empty the garbage, and teach the children and who would decide that? Would the world be Communist? Would the Third World control it? Would all whites die? Would all sex perverts die? Who would run the prisons—would there be prisons? Endless questions like these were raised by the Weathermen, but we didn't have answers."[18] Similarly, jihadists today are political ignoramuses because their leaders have abdicated teaching or leadership of any sort. They champion "militant jihad irrespective of whether a recruit has received religious training or not." As Nelly Lahoud points out, "They have introduced a shortcut, declaring that Muslims cannot afford the luxury of an extensive religious education before embracing jihad..."[19] Islam scholars like Musa al-Qarni have long been critical of jihadist thinkers not only for their tactics, but also for offering "no religious education."[20] Task cohesion requires that leaders teach the agenda to inform the rank-and-file of the goal and dissuade newcomers uninterested in working towards it.

Teaching the Means

Along with their political ends, the smart leader educates the rank-and-file about the optimal means to attain them. Tactical education is critical for leaders to impose restraint even in nonviolent social movements. The social philosopher Richard Gregg relays this story about Martin Luther King: "On the night of January 30, 1956, a bomb was thrown on the porch of Dr. King's house...A crowd of angry Negroes gathered, but Dr. King pleaded with them not to be violent or angry, and they obeyed him."[21] The American historian David Garrow tells a related story of the American civil rights icon: "King convinced Richard 'Peanut' Tidwell, the twenty-year-old leader of the Roman Saints, to accept his point. Tidwell joined him in telling the other gang chieftains to give King's proposals a try, and the discussion went back and forth...King's arguments won over more and more of the gang leaders. Finally, at 3:00 or 4:00 a.m. the entire group agreed to tell their members to avoid further violence."[22] Mahatma Gandhi also knew he had to teach the nonviolence of *Satyagraha* to members lest they escalate uncontrollably against the British.[23] In their seminal book on nonviolent protest, the social movement theorists Gene Sharp and Marina Finkelstein point out that leadership "serves a very important role, especially where knowledge and understanding of the principles and practice of nonviolence action are not widespread and deep among the general population."[24] Tactical education remains vital when nonviolent groups transition into violent ones to restrain members from going too far and shooting the cause in the foot.

The Animal Liberation Front (ALF) and Earth Liberation Front (ELF) are generally thought of as "leaderless" resistance groups, but the leaders actually implore participants to use violence sparingly. Both the ALF and ELF credos instruct them "To take all necessary precautions against harming life."[25] To cultivate task cohesion in the ranks, smart militant leaders instill in members the perils of indiscriminate violence. The most successful rebels have taught their foot-soldiers to court the people. Mao impressed on Red soldiers to "keep the closest possible relations with the common people" not only by sparing them, but by replacing the door when leaving a house (doors were sometimes attached and used as beds), paying for items, and establishing latrines away from residential areas.[26] Similarly, Vietnamese general, Vo Nguyen Giap, offered this advice to his army: "In contacts with the people, follow these three recommendations: To respect the people, to help the people, to defend the people...Our army has devoted great attention to the strengthening and development of friendship with the people."[27] Che Guevara also wasn't content merely to know the political costs of harming the population; he told his men to "avoid useless acts of terrorism."[28] His fighting manual stresses that "terrorism is of negative value, that it by no means produces the desired effects, that it can turn a people against a revolutionary movement."[29] In the *Minimanual of the Urban Guerrilla*, the Marxist Brazilian revolutionary, Carlos Marighella, likewise cautions his foot-soldiers not to "attack indiscriminately without distinguishing between the exploiters and the exploited."[30] In his detailed case study of the Irgun, Bruce Hoffman notes that the leadership always advised "targeting the physical manifestations of British rule while deliberately avoiding the needless infliction of blood-shed."[31] It wasn't enough for Begin to grasp the costs of terrorism; he made sure that his subordinates knew them, too.

Begin's targeting guidelines were quite specific. He instructed operatives to attack British-occupied infrastructure at night "when the offices were likely to be empty and there were few passerby [sic] on the streets."[32] For example, the Irgun bombed offices in the Department of Taxation and Finance—the principal organs of the occupation responsible for revenue collection—between 8:30 p.m. and 10:30 p.m., outside normal business hours.[33] Begin was so committed to limiting gratuitous British deaths that he commanded his operatives to attack targets during Shabbat "because only at that time was the neighborhood comparatively unfrequented by civilians."[34] The Irgun leader also pioneered the practice of issuing pre-attack warnings to spare civilians. As U.S. Army intelligence noted from Cairo about the group, "The distinguishing feature...is the Irgun Zvai Leumi is waging a general war against the government and at all times took special care not to cause damage or injury to persons, going so far as to post warnings on the mined places advising all to stay away."[35] Commanders knew to post signs of impending attacks, like the posters printed in Hebrew, Arabic, and English on the walls of Jaffa and Jerusalem that ominously read: "WARNING! The Government of Oppression should WITHOUT ANY DELAY evacuate children, women, civilian persons and officials from all its offices, buildings, dwelling places etc. throughout the country. The civilian population, Hebrews, Arabs, and others are asked, for their own sake, to abstain from now until the warning is recalled from visiting or nearing Government offices, etc. YOU HAVE BEEN WARNED!"[36] As I mentioned in Chapter 6, these warnings to avert "collateral damage" didn't always work, like the one tragically ignored on that fateful summer day in 1946 to evacuate the King David Hotel.

ANC leaders like Mandela credit the Irgun for their restraint and executed it even better.[37] The leadership worked hard at tactical education, as one ANC member recalled: "We went to classes where we learned to use various weapons, explosives, etc., and studied the tactics of guerrilla and mobile warfare... In the afternoons, I returned to the house for classes on military tactics taught by members trained elsewhere. This way we exchanged views on various military practices and strategies from different countries."[38] The ANC started off nonviolent, but determined in the 1960s that ending the apartheid required an armed struggle. The main tactical advice from this point forth was for members to avoid "terrorist acts."[39] Fighting manuals stressed that violence should be directed only against "military forces" and other "government stooges."[40] After the Wimpy bar misfire, senior members went to the front lines with a message: "No more indiscriminate bombing of civilians."[41] The rank-and-file sometimes made mistakes, but understood their targeting instructions. When a truculent fighter proposed bombing a building in the vicinity of children, for example, another member explained how such risks ran counter to policy: "His attitude was in direct contradiction to what I understood...policy to be, namely that every effort would be taken to prevent human injury, and that attacks would be launched against symbols of apartheid, such as government buildings and military installations, as well as power plants and communication lines."[42] Indeed, members had been explicitly told by higher-ups, "There were children playing outside until well after dark, and we had to wait until they had gone home before placing the device."[43] ANC leaders were particularly impressed by the Irgun strategy of issuing pre-attack warnings.[44] As with the Irgun, this operational advice from the ANC leadership was effective but not foolproof, like the errant bomb at a train station on July 24, 1964, that hurt twenty-three white commuters who didn't get the message.[45] The IRA leadership shared similar targeting instructions with their members. Official policy was to avoid civilians even if they formerly worked for the British army.[46] Like in the Irgun and ANC, IRA leaders taught members to cancel operations that might kill civilians or at least attack in a manner that limited the chances.[47] Even when the IRA struck civilian targets, its habit of issuing warnings meant that most attacks didn't kill anyone.[48] This procedure wasn't unerring for the IRA or its offshoots either.[49] In his autobiography, for example, Eamon Collins recounts how even though his unit phoned in a warning to a customs station office shouting "Bomb! Bomb! There's a bomb downstairs!" an eleven-year-old boy was killed before the police could clear the area.[50]

Smart leaders of insurgent groups enshrine their targeting guidelines in a "code of conduct" for fighters to follow. Yoweri Kaguta Museveni of the National Resistance Army prevailed in the Ugandan Bush War of the 1980s partly because of his clear rules of engagement. The code of conduct read: "Never abuse, insult, shout at, or beat any member of the public. Never take anything in the form of money or property from any members of the public ... Never kill any member of the public or any captured prisoners, as the guns should only be reserved for armed enemies or opponents."[51] Fighters attest that "NRA's leaders were concerned with protecting the civilian population ... You must go out of your way to help and support the local population by providing physical protection, water, food, and any other assistance local people required."[52] The Free Syrian Army's official code of conduct is almost entirely devoted to the proscription of civilian attacks. Article 1 states: "I will direct my weapons exclusively against Assad aggressors ... I will respect human rights in accordance with our legal principles, our tolerant religious principles, and the international laws governing human rights—the very human rights for which we struggle today and which we intend to implement in the future Syria." Article 8 affirms: "I pledge not to use my weapon against activists or civilians, whether or not I agree with them; and I pledge to not use my weapon against any other Syrian citizen. I pledge to limit my use of weapons to the defense of our people and myself in facing the criminal [Assad] regime." Article 9 likewise forbids sectarian violence: "I pledge not to exercise reprisals on the basis of ethnicity, sect, religion, or any other basis, and to refrain from any abusive practices, in word or in deed, against any component of the Syrian people."[53] FSA members haven't always followed these targeting instructions, like when they ransacked and shelled pro-government towns in West Aleppo.[54] But partisans debate how much the FSA leadership condones such indiscriminate violence—not its official stance against it.[55] The same is true of the Afghan Taliban whose top-level leadership has consistently warned fighters not to target civilians since the first Layha of 2006.[56]

After the Zarqawi-led debacle in Iraq, Bin Laden and his deputies tried to teach Al Qaeda members about the fallout from indiscriminate violence. Letters from Bin Laden went out to commanders: "I call [the mujahedeen] to issue orders to all battalions and companies fighting in the field to prevent explosions and methods that kill generally and indiscriminately...incidents of this sort, if ever found, should be very, very rare."[57] Over time, Al Qaeda leaders impressed on subordinates the value of more selective targeting: "Committees should look into every proposed operation separately to guarantee its precise implementation...They should not attempt to hit any suspicious target or something that might cause controversy or discussion or raise debates, unless it's fully checked, 100-percent sure, and reassured that it's proper and permissible...The mujahidin leadership should advise the suicide bomber of this and warn them from being tricked or being sent to questionable suspicious targets of which they were not advised."[58] Zawahiri urged Muslims to "embrace jihad but avoid indiscriminate slaughter" in his statement marking the ninth anniversary of the September 11 attacks.[59]

Al-Aqsa Martyrs' Brigades leader, Marwan Barghouti, also instructs operatives against "the targeting of civilians inside Israel, our future neighbor."[60] The PKK leader, Murat Karayilan, directs his forces to engage "military targets" but "not harm civilians."[61] Doku Umarov, the former head of the Caucasus Emirate, wasn't known for a kind heart, but even he sometimes admonished his fighters "to focus their efforts on attacking law enforcement agencies, the military, the security services, state officials" while "protect[ing] the civilian population."[62] A former British Neo-Nazi offers an inside perspective on how the leadership educated the rank-and-file about the perils of over-escalation: "While rank-and-file members of the more mainstream groups like the National Front and the British National Party were often keen to pursue more extreme methods, the neo-Nazi hierarchy has always resisted a foray into full-scale terrorism...There were always elements from the grass roots (sic) who wanted us to move toward terrorism, but these demands were always resisted by the leadership. It wasn't that they necessarily had any moral problem with it, more that they were worried the tactics would backfire."[63] In sum, numerous militant leaders throughout history have advised the rank-and-file to spare the population, if not for moral reasons then for strategic ones.

Stupid leaders, by contrast, give no targeting guidance. In Renamo, for instance, there was "no clear, specified code of conduct" against harming civilians. The leadership "did not share expectations [with militants] about how they should behave, which behaviors were permitted and which prohibited."[64] Even stupider leaders actively encourage members to strike civilians. As we've seen, GIA leaders throughout the 1990s from Abdelhaq Layada to Djamel Zitouni to Antar Zouabri issued a series of fatwas calling for an ever-expanding list of targets until almost everyone in Algerian society was sentenced to death as an infidel.[65] Islamic State leaders didn't just read "Management of Savagery" for fun; it became the bible for operatives to follow. In September 2014, for example, ISIS spokesman Abu Mohammed al-Adnani issued an audiotaped appeal for soldiers of the caliphate to attack people all over the world with any means at their disposal: "If you are not able to find an IED [improvised explosive device] or a bullet, single out the disbelieving American, Frenchman, or any of their allies. Smash his head with a rock, or slaughter him with a knife, or run him over with your car, or throw him down from a high place, or choke him, or poison him."[66] In July 2017, Islamic State leaders released an e-book to its Turkish franchise over the Telegram instant messaging service beloved by the group for its end-to-end encryption. The sixty-six-page manual with 174 illustrations and seven charts calls upon the *umma* to step up attacks against soft targets. Detailed instructions call for burning parked cars, setting forest fires, creating traps for highway accidents, and the growing art of car-ramming pedestrians. The manual reassures readers that attackers shouldn't dwell on distinguishing between military and civilian targets or even which country to terrorize. Killing people in Europe and U.S. is probably best, but attacks in Turkey might be easier to pull off, so perhaps start there.[67] Even locally, ISIS fighters have been given what one described as a "green light" to do whatever they please to young girls and other civilians.[68]

Testing Education and Task Cohesion in Militant Groups

The instructions of leaders seem to affect task cohesion in militant groups, particularly their tendency to strike civilians. Members of Al Qaeda, the IRA, KKK, the Malayan National Liberation Army, and other groups limited their violence against civilians after the leaders acknowledged the political costs. Conversely, members of other groups have historically ramped up their attacks on civilians in line with targeting appeals from the top. GIA members turned against Algerian society in response to fatwas, especially Zouabri's in August 1996, declaring apostate any Algerian who refuses to fight the government. Shortly afterward, GIA bandits massacred with knives hundreds of women and children in Rais, Bentahla, and Relizane among other villages.[69] According to Al Qaeda leader, Atiyya Abd al-Rahman, GIA chiefs "destroyed themselves with their own hands, with their lack of reason, delusions, their ignoring of people, their alienation of them through oppression, deviance, and severity, coupled with a lack of kindness, sympathy, and friendliness . . . They defeated themselves."[70] Similarly, PKK members appear to have followed the targeting appeals of Abdullah Ocalan. In June 1993, for example, he ordered operatives to launch attacks against tourist sites throughout Turkey. Ocalan boasted: "We are going to wage an all-out war against it [Turkey] until it agrees to negotiate . . . We will hit economic and tourist interests throughout Turkey."[71] This announcement was immediately followed by attacks in Istanbul and other tourist areas around the country.[72] The leaders claimed credit in their March 1994 conference: "All economic, political, military, social and cultural organizations, institutions, formations—and those who serve in them—have become targets. The entire country has become the battlefield."[73] The ISIS leadership's targeting appeals have also been inspirational, with operatives throughout the world attributing their massacres to Abu Bakr al-Baghdadi himself.[74] In these ways, the stance of leaders toward terrorism clearly seems to affect whether members perpetrate them.

To assess this relationship more scientifically, let's return to the MAROB dataset of 118 organizations operating in sixteen Middle East and North Africa countries. This dataset includes variables on whether the leader was known to sanction civilian attacks and whether his members committed them. This information can help us evaluate whether the instructions of the leadership significantly affect the behavior of its subordinates. Because MAROB is structured as time-series data, it can account for situations in which leaders have adopted inconsistent positions toward civilian targeting over time. If my argument is correct, we should find that historically members of militant groups have been less likely to engage in terrorism when their leaders were known to oppose this politically counterproductive practice. And this relationship should hold even after taking into account many other factors like the ideology of the group, its capability, and that of the target country. It turns out that militant groups are indeed significantly less likely to attack civilians when the leaders take a stance against it. Table A.12 shows the odds of this relationship resulting from chance are less than one in a thousand even after controlling for such factors as the membership size of the group, its territorial control, popular support, ideological orientation, and other potential confounds. The substantive impact of the leader on target selection is very large; at least in the MENA region, a militant group has been about fifty times more likely to attack civilians when the leader was known to favor this target choice. Both qualitative and quantitative evidence thus indicate that a leader's stance toward terrorism strongly affects the proclivity of members to use it.

In *Inside Rebellion*, the political scientist, Jeremy Weinstein, purports to "highlight the importance of structure over agency" in accounting for why some militant groups target civilians.[75] But my analysis suggests that militant leaders wield considerable influence over the targeting choices of lower-level members and hence the prospects of victory. The successful leader may not give a damn about protecting innocent life. But it doesn't matter what's in his heart. It doesn't even matter what's in his head if his subordinates remain in the dark. Smart leaders know not only the political costs of attacking civilians, but also to educate the rank-and-file about them. Such education is essential for rebels to cultivate task cohesion in the ranks, overcome the principal-agent problem in the group, and triumph, especially if they keep reading this book. In Chapter 9, we'll see how smart leaders ensure members heed their tactical advice.

Notes

1. Richardson July 26, 2013.
2. MaCoun et al. 2006, 646.
3. Lichbach 1998, 167.
4. Mullen and Copper 1994, 210.
5. Gamson 1975, 91.
6. Duka et al. 1974.
7. Cambanis 2011, 7.
8. Hamzeh 2004, 76. These standards have been relaxed. Szakola, September 2, 2016.
9. Weinstein 2006, 371.
10. Clausewitz 2017.
11. Barnett 1972, 13.
12. Gaddis 2005.
13. Duka et al. 1974, 8–9.
14. Weinstein 2006, 145.
15. Ibid.
16. Stanton 2016, 251.
17. Kaplan and Marshall 1996, 85.
18. Stern 2007, 96.
19. Bhasin, May 26 2011
20. Ibid. 19.
21. Gregg 1966, 39.
22. Garrow 1986, 496.
23. Ghandi 1948, 470.
24. Sharp and Finkelstein 1973.
25. Brown 2017.
26. Boot 2013, 333.
27. Giap 2001, 62–3, 132.
28. Sinclair 1970, 33.
29. Guevara 2002, 116.
30. Marighella 2011.
31. Hoffman 2016, 127
32. Ibid.
33. Ibid.
34. Begin 177, 96.
35. Hoffman 2016, 143.
36. Ibid. 2016, 203.
37. Mandela 1964.
38. Duka et al. 1974.
39. Slovo 1997, 178.
40. Turok 2003, 280.
41. O'Malley 2007, 237.
42. Kathrada 2004, 143.
43. Ibid.

44. Lewin 2011, 114.
45. Ibid. 109, 111, 128.
46. Collins and McGovern 1998, 107.
47. Donnelly 2010, 62–4.
48. Brown 2017, 7.
49. Moriarty March 3, 1998.
50. Collins and McGovern 1998.
51. Weinstein 2006, 371.
52. Wilson 1989, 141.
53. "The Armed Conflict in Syria."
54. Wagner.
55. *The New Arab*, 2 November 2016.
56. I go into the Taliban more in Chapter 11.
57. "How Al-Qaeda In Iraq (AQI) Has Transformed," 5 March 2016.
58. Ibid.
59. Gerges 2016, 125.
60. Barghouti 2002, A19.
61. Mavioglu, October 28, 2010.
62. Dzutsev 2012.
63. "We're at War," 26 April 1999.
64. Weinstein 2006, 147.
65. Tawil 2011, 125.
66. Bazzi, July 27, 2016.
67. Yayla, April 26, 1999.
68. "Captures ISIS Militant," 17 February 2017.
69. "Bentalha," April 8, 1999.
70. Al-Rahman, December 12, 2005.
71. Yassine, June 8, 1993.
72. Stanton 2016, 218.
73. Gunter 1997, 49.
74. Clarke and Gartenstein-Ross 2016.
75. Weinstein 2006, 21.

9

The Structure of Success

Members take their targeting cues from the top. Regardless of their capability or ideology, militant groups are far less likely to perpetrate terrorism when the leaders stand against it. For the rebel, this is welcome news. It means he can personally influence the quality of tactics and hence likelihood of victory. But let's not overstate the value of words alone. Members sometimes attack civilians even when the leaders are known to reject this counterproductive behavior. Taking a stance against terrorism improves targeting decisions in the aggregate, but hardly guarantees tactical restraint. The political scientist Joseph Nye notes in his book on leadership, "You cannot lead if you do not have power."[1] Naturally, the militant leader must possess enough organizational clout to ensure that members faithfully execute his targeting preferences. For the militant leader, the key to task cohesion is centralizing the organization.

A basic premise of organizational theory is that group structure affects the locus of decision-making. The more centralized an organization, the less autonomy is delegated to subordinates.[2] For this reason, group structure is a standard proxy for leadership control in many types of organizations.[3] A more centralized militant group reduces civilian attacks in three main ways. Centralizing the organization helps the leader to (1) communicate his tactical instructions to the rank-and-file, (2) discipline wayward members for attacking civilians, and (3) vet out high-risk recruits prone to subverting the cause with terrorism. This chapter explains the benefits to centralizing before quantifying them statistically.

Communications

Leaders have an easier time conveying their preferences to members in more centralized organizations, says a large body of research in management and sociology.[4] Every organization suffers principal–agent problems because subordinates are never as capable and committed as leaders desire. Communication is therefore always imperative for getting subordinates to do what leaders want. But communication is especially important in social movements given their unusually high degree of internal heterogeneity. As we saw in Chapter 7, the leaders of social movements are nothing like the rank-and-file. There's generally a huge disparity in formal education, combat experience, and aptitude. Without communications from the leader, low-level members can't possibly be expected to meet his expectations. As one study on social movements notes in the introduction: "If leadership has one predominant feature, it is that leaders communicate...To lead is above all to communicate."[5] Saul Alinsky affirms: "Communication is fundamental...One can lack any of the qualities of an organizer—with one exception—and still be effective and successful. That exception is the art of communication. It does not matter what you know about anything if you cannot communicate to your people."[6] Martin Luther King also believed that the most important attribute for leadership is "the art of persuasive communication."[7] According to network scientists like Duncan Watts at Microsoft, decentralized social units "tend to be less efficient information providers."[8] Gamson finds in his study of American volunteer groups from 1800 to 1945 that more centralized ones had higher success rates, perhaps due to the communication benefits.[9]

In militant groups, centralization is indispensable to the leader in supplying tactical guidance to members lest they strike the wrong targets. Based on his experience in the Polish underground during the Second World War, Janusz Zawodny concluded: "The broader the basis of acquaintance and knowledge of identities among the members [through centralization], the more efficient are the channels of communication and the better the control exercised by command over the membership."[10] Hezbollah is a prime example of a centralized group that provides clear instructions to fighters about which targets to engage. In his database of militant group structures, the political scientist, Joshua Kilberg, codes Hezbollah as possessing "the most hierarchical of all the structures," as "its command structure is dense, elaborate and centrally controlled."[11] Such centralization has enabled the longtime secretary-general, Hassan Nasrallah, to wield considerable influence over members. Ahmad Nizar Hamzeh writes, "There is no doubt that Nasrallah... [is] the central actor in almost all of Hezbollah's political and military decision making."[12] According to Matthew Levitt, military operations must be "cleared first by the organization's highest leadership command."[13] Magnus Ranstorp explains that such guidance has ensured its acts of resistance "coincided with the collective interest of the organization as a whole."[14] Nasrallah's hand was on display in May 2000 when the Israel Defense Forces withdrew from southern Lebanon. The IDF left behind thousands of collaborators, including men who had allegedly roughed up Hezbollah fighters at the behest of the Israelis. Unsurprisingly, many Hezbollah fighters were baying for blood. But Nasrallah ordered his followers not to touch the collaborators, leaving the judgement to Lebanese courts.[15] In this way, he overrode the powerful vengeance impulse in the ranks. Hezbollah is so centralized that Imad Mughniyeh and other senior leaders allegedly watched the marine barracks attacks in 1983 through binoculars from a perch atop a nearby building.[16] In the last chapter, we saw how leaders of the Irgun, African National Congress, and IRA also imbued members with the sense to abstain from indiscriminate violence. Notably, each of these organizations was highly centralized, facilitating this guidance from the top.

Conversely, the political scientists Mette Eilstrup-Sangiovanni and Calvert Jones point out, "History provides numerous examples of networks that fail in their missions due to inefficient communication and information sharing."[17] Indeed, leaders have historically struggled to communicate their targeting instructions in more decentralized groups. Pye alludes to this difficulty within the Malayan Communist Party: "An important reason why the MCP was unable to organize militarily significant guerrilla operations lay in the lack of an efficient and rapid system of communication. The necessity of depending entirely upon couriers and prearranged meetings in the jungle meant that the actions of the separate MRLA units could not be coordinated or given adequate centralized direction. The crude methods of communication... furthered the development of terrorism."[18] An autobiography of a Huks leader acknowledges a similar dynamic in the ranks: "I [have] admitted that errors were made and that innocent people died. Invariably, such tragic mistakes were the result of unauthorized action by squadrons operating far from our general headquarters."[19] Cut off from the leadership, "units that had been assigned to distant locations became loose and inefficient... impulsive and impatient when they were isolated from the central command."[20] The same fate befell the Weather Underground in 1970 when the radical left-wing organization split into factions in San Francisco, Chicago–Detroit, and New York. The decentralization of the movement eroded communications and with them the quality of resistance. Audrey Cronin notes how "lacking central direction... without the knowledge of the others" members became unmoored, accidentally bombing themselves into organizational ruin.[21] Communication problems also contributed to the demise of the loosely organized network of FLQ militants. The October 1970 crisis broke out when a couple of cells failed to coordinate their strategies. The Liberation cell kidnapped a British diplomat, while the Chernier cell kidnapped the Quebec labor minister, Pierre Laporte. After the Liberation cell publicly announced that the FLQ would release the hostages only if its demands were met, the Chernier cell ended up killing Laporte, destroying local support for the group. As one account notes, "The loose network structure and lack of central authority made reliable communication and information sharing difficult."[22] Similarly, a Central Committee member in the People's Movement for the Liberation of Angola complained about how decentralization disrupted communications from the leadership, "Many problems in the areas of... communication... emerged or were exacerbated as a result of the top leadership and headquarters being located outside Angola." Only when the leaders "work more permanently among the people" are they able to offer "more clearly defined political content."[23] Laqueur posits, "The more remote the headquarters from the scene of action, the less complete its knowledge of current events, the more tenuous its contacts with its own men."[24] Bin Laden certainly understood the difficulty of contacting his men from afar, penning frustrated letters through surrogates to his braindead associate in Iraq. As we saw in Chapter 6, the Al Qaeda founder struggled to convey to Zarqawi the imperative of civilian restraint. West Point researchers noted, "The tone in several letters authored by Bin Laden makes it clear that... the affiliates either did not consult with Bin Laden or were not prepared to follow his directives."[25] In one of the letters, Atiyah practically begs Zarqawi not to make any more hasty operational decisions without consulting the leadership in Afghanistan and Pakistan.[26] Try as he might to communicate from his Abbottabad compound the perils of indiscriminate violence, Bin Laden couldn't get the message across to fighters in distant lands.

Discipline

The Strategic Model of Terrorism posits that members attack civilians at the behest of the leadership to advance its political goals. But the history of asymmetric conflict is loaded with examples of militant leaders disciplining fighters for this self-defeating behavior. Che laid down the hammer: "When this discipline is violated, it is necessary always to punish the offender, whatever his rank, and to punish him drastically in a way that hurts."[27] Fearful civilian attacks would backfire on the Zionist goal of evicting the British from Palestine, the Haganah was known to "punish the culprits... considered responsible for terrorist outrages."[28] As David Ben-Gurion put it: "We cannot fight terrorism by condemnation alone. For people whose only argument is dynamite, persuasion is useless. We need drastic action to wipe out terrorism."[29] Jewish Agency officials boasted that intel supplied to the police accounted for 95 percent of all terrorist arrests.[30] Palestinian leaders have also disciplined terrorists. In March 1996, for instance, Yasser Arafat ordered Palestinian Authority security forces to execute a sweeping campaign against Hamas, arresting 1,200 terrorists whose extremism compromised the Palestinian cause.[31]

The punishment is often severe. Some of those captives would later charge that PA officers abused them.[32] For hurting villagers in the Mau Mau insurrection, a member got "twenty-five strokes on his buttocks," with some regional leaders treating rape as a capital offense.[33] Rape was also punishable by death in the Communist guerrilla army in the Philippines.[34] The leadership meted out other stiff sentences for "crimes against the people" from indeterminate periods of hard labor to expulsion from the group. Huks members who killed "innocent people" were "court-marshaled and punished for committing crimes against the people."[35] When caught terrorizing civilians, members of Uganda's National Resistance Army had to face public trials before elders who "administered punishment brutally and quickly to underscore the seriousness of indiscipline."[36] PKK leaders handed out a twenty-four-year prison sentence in 2010 to a small cell that attacked civilians in Turkey's Batman province.[37] Leaders of the New People's Army in the Philippines dealt out a variety of "disciplinary actions" to disobedient rebels in 2012 and 2014 for incidents against civilians.[38] In 2015, Nusra chiefs made members of the Al Qaeda affiliate stand trial before an Islamic court in Idlib for slaughtering twenty Druze villagers.[39] In other cases, the punishment isn't as dire. The IRA leadership disbanded the Fermanagh unit in 1989 after operatives committed sectarian violence against the Protestant population.[40] Prachanda suspended the Nepalese attackers behind a botched 2005 bus bombing that accidentally killed thirty-eight people.[41] These punishments are hardly cherry-picked cases from history. In a sample of 108 terrorist autobiographies, the political scientist Jacob Shapiro found that 44 percent feature examples of leaders disciplining subordinates for committing tactical and other transgressions.[42]

Why do leaders discipline members for attacking civilian targets? Research suggests that punishment lowers the incidence of this politically disastrous behavior. In Sierra Leone, for example, militant groups that were punished by their leaders for attacking civilians were significantly less likely to do it.[43] Without fear of punishment, members are free to pursue their personal interests at the expense of the group. A centralized organization is essential for leaders to make terrorism as unprofitable for its perpetrators as it is for the political cause. Centralization facilitates discipline in the ranks, as theorists have known since at least the seventeenth century with the publication of Thomas Hobbes' *Leviathan*. The Hungarian Communist revolutionary Bela Kun published a manifesto in the early 1920s entitled "Discipline and Centralized Leadership." As the title suggests, Kun argued that "discipline comes not only from the masses, but mainly from the leaders, and it requires therefore...well-organized cadres."[44] A study of nineteenth and twentieth century American protest groups determined, "Centralization of power is an organizational device for handling the problem of internal division and providing unity of command...A centralized, bureaucratic group that escapes factional splits is highly likely to be successful."[45] Drake remarks in his study of terrorist target selection, "The 1950s Greek-Cypriot group EOKA was highly centralized, with its commander, George Grivas, exercising strict—though not total—control over all aspects of operations..."[46] By contrast, leaders in highly decentralized organizations struggle to punish fighters. Throughout the 1990s, the founder of Jemaah Islamiyah, Abdullah Sungkar, ruled with an iron fist the highly centralized Southeast-Asian terrorist group. With his death in 1999, the group decentralized and lacked an effective enforcer to discipline the increasingly terrorism-prone factions.[47] A former CIA operations officer points out that there's "no way to impose discipline" on today's jihadists even if the leadership wanted because they're so geographically dispersed.[48] Centralization thus helps leaders to restrain members not only in communicating to them which targets to avoid, but for punishing those who don't.

Vetting

Centralization also helps the leader to weed out recruits who risk spoiling the cause with terrorism. Vetting is undervalued for militant groups and social movements more generally. Nearly all scholars prioritize the quantity of participants over their quality by focusing exclusively on the benefits of larger membership rosters. In *Power of Numbers*, for example, DeNardo stresses: "Oppositional movements appear to share at least one thing in common. Regardless of the political context, there always seem to be power in numbers."[49] The community activist, Si Kahn, espouses a similar sentiment, "The power of a lot of people working together is enough to make changes where one person can do very little."[50] The economist Mancur Olson wrote, "A group of people small enough to engage readily in collective action usually won't have enough strength to overthrow any effective government."[51] The political scientists Erica Chenoweth and Maria Stephan confirm that civil resistance movements have historically fared better with larger memberships.[52] All else being equal, social movements seemingly gain power as the number of members rises.

The problem is that not all members are equal. In fact, there's often an inverse relationship between the quantity and quality of members in any organization, as university admissions officers know all too well. Striking the right balance between quantity and quality is an open question; what's not is the advisability of an open-door policy. An open-door policy exacerbates principal–agent problems by letting in talentless, even nefarious recruits while increasing the monitoring costs of screening out bad apples from the ranks. The political economist Elinor Ostrom uses the term "span-of-control problems" to argue that "the cost of monitoring increases with the size and diversity of a firm or a state."[53] As the political scientist Mark Lichbach recognized, "Given that monitoring costs are obviously lower in small groups, dissident entrepreneurs might well prefer smaller to larger organizations."[54] Sharp and Finkelstein also warn protest leaders, "Very careful consideration must constantly be given to the relationship between the numbers participating in the conflict and the quality of their participation . . . Large numbers may even be a disadvantage."[55] The Quaker activist George Russell Lakey also advised protest leaders against lax screening because "quality may be more important, even at the cost of numbers."[56] Gandhi, too, attached "the highest importance to quality irrespective almost of quantity."[57] So did Nusra. "We pay a great deal of attention to the individual fighter, we are concerned with quality, not quantity," an Aleppo-based leader boasted in a 2012 interview with *Time*.[58] To this end, Nusra practiced *tazkiyah*—a process requiring recruits to be vouched for by a veteran and then undergo months of training.[59]

Militant groups have paid dearly for "scaling" too quickly by prioritizing the quantity of fighters over their quality. Take the Tupamaros in Uruguay. They began operations with fifty members in 1966. Within five years, the membership roster had swollen to 3,000. On paper, this rapid growth must have seemed like progress. In reality, though, the acceptance of so many new members with questionable skill and dedication led to unrestrained outbursts of violence, permanently destroying organizational support.[60] Aum Shinrikyo, the doomsday cult notorious for its 1995 sarin gas attack in Tokyo, also experienced growing pains. Aum counted up to 40,000 members worldwide at its peak in late 1990s. But with more members came less loyalty, leading to endless disputes with the leadership and ultimately organizational fragmentation.[61] Militia groups in the U.S are notorious for taking almost anyone who volunteers.[62] Without proper vetting, these groups have been repositories for duds and informers, like the Oregon militia that disbanded in May 1995.[63] Some militias have been more selective with admissions, like the 52nd Missouri Militia that considered only individuals with a background in electronic countermeasures, intelligence surveying, and mapmaking. Applicants with those skills still had to get a recommendation from a veteran 52nd member.[64] According to law enforcement, militia groups in the United States have learned from getting burned so many times to exercise more vigilance about whom they admit.[65]

A degree of centralization is required for quality control. The political scientists Harold Lasswell and Abraham Kaplan observed that in centralized groups leaders can restrict "the permeability of a group," that is, "the ease with which a person can become a participant."[66] Centralized groups suffer a lower risk of what economists call "adverse selection" and "moral hazard" problems. With superior centralization, leaders can inspect applicants more thoroughly and reject unpromising ones posing as helpers. And by controlling the membership size, leaders can keep better tabs on those who threaten the cause. Della Porta explains how centralization aided the Red Brigades with quality control: "Decision-making power was centralized . . . [O]nly individuals who were able to pass a long screening test that evaluated military courage and fidelity to the organization were accepted. The rules on centralization and vertical hierarchy were followed without deviation."[67] Not surprisingly, centralized groups with rigorous vetting have been among the most successful in modern history. The Haganah leadership was picky about its membership composition; it "sought to recruit only trained military experts."[68] In the ANC, too, "Recruiting was done with extreme care."[69] The Irish Republican Army in the war of independence also relied on a tight-knit organizational structure to ensure competent, committed members. One historical account describes the value of leadership control, "The Irish Republican Army was [kept] small ... It is always easier to work to a detailed plan with small picked forces which can be implicitly trusted and controlled rather than with the spontaneously moving masses of a revolutionary populace ... It was undoubtedly wise tactics to entrust militant operations to a picked body of men ... The actual organization ... seems to have combined a high and successful degree of centralization."[70] Its Republican successor also maintained a centralized structure to select reliable members. The IRA leadership turned away countless young men who volunteered on a whim, especially after British paratroopers killed Catholics in agonizing events like Bloody Sunday.[71] When his uncle was killed on that tragic day in 1972, the leadership advised a young man against reflexively enlisting, "I should take time to make up my mind so that no one could ever accuse me of letting emotions cloud my judgment, so no one could say 'you're joining the IRA just because of Bloody Sunday.'"[72] Perhaps no major militant group has been more selective in its recruitment practices than Hezbollah. Historically, new recruits have been subjected to two stages of assessment in order to become party members. The first stage is called the reinforcement (ta'bia) and it lasts at least a year. During this time, recruits are taught all about Hezbollah's leaders and how to serve their political agenda. Recruits who demonstrate commitment, endurance, and loyalty are accepted into the party's membership. Senior officials keep a record on each member both before and after joining the party. Letters of recommendation from Hezbollah's clerics or other trusted ulama may be considered as a substitute for the ta'bia. But a security clearance from the party's security apparatus is still required. Upon passing the ta'bia, recruits undergo the second year-long stage, discipline (Intizam), where they undergo rigorous physical exercise and military training.[73] These standards have been relaxed over the years, especially during the urgent campaign to repel Salafi jihadists in the so-called Syrian civil war. But throughout, Hezbollah leaders have consistently leveraged its centralized organizational structure to distill high-quality fighters, earning a reputation for ruthless efficiency.[74]

Some militant groups, by contrast, have been too decentralized to vet. Bin Laden, for example, had little control over the nameless men who enlisted in affiliates under the Al Qaeda banner. As we'll see in the Chapter 10, adding affiliates is a surefire way to ramp up membership size. But doing so requires the principal to cede control to agents and thereby erode the quality of organizational violence. The ISIS leadership takes pride in its decentralized structure, ensuring a lower quality fighter than in its more centralized rivals.[75] ISIS is too diffuse to worry about admissions standards. There are no SAT tests, transcripts, or letters of recommendation; ISIS university accepts even the most worthless applicant. And that's exactly what it gets, especially its international cadre of lone wolves who are the most decentralized of them all.

You don't get more decentralized and worthless than Man Haron Monis, the ISIS-inspired lone wolf responsible for the so-called Sydney siege. On a sunny morning in December 2014, the Iranian-born Australian citizen decided to take hostages at the Lindt Chocolate Café and then shoot at them. Research into the perpetrator revealed no real ties to ISIS, but much evidence that he suffered from mental illness. ISIS leaders couldn't care less who he was; they had never heard of him. But that didn't stop Monis from promoting himself as an undercover Iranian intelligence agent, an expert in black magic, a Muslim cleric of both Sunni and Shiite sects, and ultimately a soldier of the caliphate. For a living, he ran a "spiritual healing" business for women which required them to submit to sexual molestation. Months before the attack, he was charged with being an accessory to the murder of his ex-wife as well as with over forty counts of sexual assault.[76] According to the Australian government, Monis claimed "a number of religious/ideological affiliations" from being secular to a Shiite Muslim to a Sunni. The government found he was not "politically motivated," but simply "experienced bouts of mental illness" for which he had long been having professional treatment.[77] At the inquest into the Lindt café siege, psychiatrist Dr. Jonathan Phillips diagnosed Monis as "a dangerous psychopath" suffering from "a complex personality disorder."[78] His background, though eclectic, was in some ways unsurprising. Lone wolves are essentially agents without a principal. By definition, a lone wolf isn't screened by the organization. Literally anyone can become one. So, the quality is naturally as low as the admissions standards. It's no surprise that statistical studies have found that lone wolves across many ideological persuasions disproportionately suffer from mental illness and also attack civilian targets.[79] Without vetting, lone wolves are often unstable and lash out against the population under the guise of radical politics.[80] Decentralized groups lack a mechanism to weed out these people and other unrestrained members who sully the cause.

In Chapter 8, I showed that militant groups are significantly less likely to strike civilians when the leader opposes this counterproductive practice. This is great news for the rebel. But it's not enough to ensure task cohesion in the ranks. He must also possess sufficient sway within the organization to impose his targeting preferences on lower-level members. In this chapter, I suggested multiple ways in which centralizing can help to promote tactical restraint.

Centralized organizations aid the leader in communicating tactical advice to the rank-and-file, disciplining unruly operatives who harm civilians, and vetting out high-risk recruits prone to terrorism. In Chapter 10, we'll quantify the benefits of centralization on militant group tactics.

Notes

1. Nye 2008, 27.
2. Galbraith 2008.
3. See, for example, Ferrell and Skinner 1988.
4. Mintzberg 1989. See Jackson 2006 on relationship between communications and militant group structure.
5. Barker et al. 2001, 7.
6. Alinsky 1971, 69, 81.
7. Grint 2000, 206.
8. Watts 2004, 157.
9. Gamson 1975, 93.
10. Zawodny 1978, 279–80.
11. Kilberg 2012, 813.
12. Hamzeh 2004, 48.
13. Levitt 2013, 33.
14. Ranstorp 1997, 64–5.
15. Cambanis 2011, 5–6.
16. Levitt 2013, 24.
17. Eilstrup-Sangiovanni and Jones 2008, 20.
18. Pye 1956, 99.
19. Taruc 1967, 23.
20. Ibid. 158.
21. Cronin 2011, 103.
22. Eilstrup-Sangiovanni and Jones 2008, 20.
23. Barnett, 1972, 5–6.
24. Laqueur 2016, 85.
25. Lahoud 2012, 13, 22.
26. Scheuer 2011, 147.
27. Guevara 2002, 113.
28. Hoffman 2016, 138.
29. Ibid. 178.
30. Ibid. 191.
31. Pearlman 2011, 140.
32. Ibid.
33. Barnett and Njama 1966, 193.
34. Taruc 1967, 30–1.
35. Ibid. 23, 88–9.
36. Weinstein 2006, 145.

37. "PKK Punished," December 20, 2010.
38. "Communist Rebels Sorry," September 7, 2012; "Philippine Rebels Apologize," March 7, 2014.
39. "Syria's al-Qaeda Affiliate Says it Regrets," June 13, 2015.
40. Bell 1979, 610.
41. *Associated Press*, June 8, 2005.
42. Shapiro 2013, 69.
43. Humphreys and Weinstein 2006, 441.
44. Kun 1923.
45. Gamson 1975, 108.
46. Drake 1998, 164.
47. Eilstrup-Sangiovanni and Jones 2008, 28.
48. Sageman 2008, 146.
49. DeNardo 1985, 35.
50. Kahn 1992, 2.
51. Olsen 1990, 13.
52. Chenoweth and Stephan 2008, 39.
53. Ostrom 1990, 222.
54. Lichbach 1998, 215.
55. Sharp and Finkelstein, 475, 498.
56. Lackey 1962, 53.
57. Dhawan 1946, 225.
58. Abouzeid, December 25, 2012.
59. This practice isn't consistently implemented. The point is that the leaders understand the tradeoffs of membership numbers and quality unlike, say, ISIS. See, for example, Nance 2016, 195.
60. Laqueur 2016, 85; Brum 2014.
61. Eilstrup-Sangiovanni and Jones 2008, 25.
62. Snow 1999, 69.
63. Kovaleski and Schmidt, May 6, 1995.
64. Snow 1999, 69.
65. Ibid. 131.
66. Laswell and Kaplan 2013, 35.
67. Della Porta 2006, 115.
68. Malet 2013, 199.
69. Turok 2003, 49.
70. Chorley 1973, 58–9.
71. Richardson 2007, 89.
72. Taylor 1999, 152.
73. Hamzeh 2004, 75–6.
74. Leung, April 8, 2003.
75. Crary, July 28, 2016.
76. "Self-styled Muslim Sheikh," November 22, 2013.
77. Commonwealth of Australia, 2015.

78. Jones, August 13, 2016.
79. Gill et al. 2014; Becker 2014.
80. Van Zuijdewijn 2015; Johnson, July 15, 2016. Salient examples include the Germanwings pilot who intentionally crashed his commercial airplane into the Alps in March 2015 and the person who tried to assassinate President Reagan to somehow impress Jodie Foster. In these cases, mental health experts concluded that Andreas Lubitz was "unfit to work" due to his suicidal tendencies, while John Warnock Hinckley was found "not guilty by reason of insanity."

10

The Benefits of Centralizing

The first part of this book revealed that certain tactical choices are better bets than others. The Strategic Model of Terrorism is a compelling theory until you test it. It turns out that groups are far more likely to triumph when they abstain from terrorism. Civilian attacks only feed the impression that the perpetrators are blood-thirsty maniacs bent on slaughtering innocent men, women, and children. Naturally, political concessions are rejected as pointless when the perpetrator is seen as an unappeasable killer. Given this perception of terrorists, it should come as no surprise that they scare off supporters and invite crushing government reprisals. The first rule for rebels is thus recognizing the costs of terrorism to the group. The second rule is to restrain its members from doing it. That's easier said than done because they're often indifferent to the political plight of the group, oblivious about how to advance it, and may even benefit by subverting it with terrorism. Leaders can minimize this principal–agent problem by educating the rank-and-file about their political ends and the optimal means to achieve them. But some members will commit terrorism anyway in defiance of the leadership. Centralization may hold the answer for cultivating task cohesion in the ranks. Last chapter, we saw how structuring the organization in this way can help leaders to communicate their tactical instructions, discipline wayward members, and vet out potential terrorists. This chapter demonstrates more scientifically the extent to which centralization prevents members from shooting the cause in the foot.

Terrorism in Decentralized Groups

MAROB can assist in quantifying the benefits of centralization on militant group tactics. This dataset contains variables on whether the 118 groups in the Mideast and North Africa were centrally organized and attacked civilians. In the analysis below, the independent variable is a binary measure of whether the group is centralized. And the dependent variable is a dichotomous measure of whether the organizations engaged in terrorism by attacking civilians. As in previous tests, civilian targets include any non-security state personnel, thus excluding military and police forces. Table A.12 displays the results of seven logistic regressions based on my study with Phil Potter in *International Organization*.[1] Model 1 is a simple bivariate test of the relationship between organizational structure and target selection. This parsimonious model maximizes the available data, while ensuring that the observed effect of group structure doesn't come from bias generated by missing data in the covariates. As Chapter 9 foretold, decentralized groups are significantly more likely to engage in terrorism by attacking civilians. In practical terms, there's a decline from an approximately 40 percent chance that a decentralized group will attack civilians to a 25 percent chance that a centralized group will do so. When they're decentralized, militant groups are thus about 15 percent more likely to target civilians. At least in Middle Eastern and North African countries, decentralized groups have been associated with more counterproductive attacks.

This preliminary finding establishes a link between civilian attacks and decentralized groups, but doesn't resolve why. It's conceivable that decentralized groups are inclined to terrorism simply because their leaders favor this tactic rather than due to a loss of organizational control. The key question for the leader is whether centralizing the group lowers the incidence of terrorism when he's smart enough to oppose this counterproductive targeting practice. Lucky for us, MAROB also includes a variable for whether the leadership sanctions civilian attacks as we discussed in Chapter 8. This information is invaluable for determining whether decentralized groups are associated with terrorism due to the tactical preferences of their leaders or weak organizational control. For the rebel, this information enables us to determine whether centralizing his organization can improve the quality of its tactical choices and thereby boost the odds of victory. Model 2 gets to the bottom of this question by testing the impact of group structure on civilian attacks conditional on whether the leadership favors them. Evidently, leaders smart enough to oppose terrorism and centralize their groups can restrain members from mucking up the cause with civilian attacks. The interaction term for organizational structure and leadership preferences is statistically significant at the 0.001 level, meaning the likelihood of this finding occurring by chance is less than 1 in a thousand.

Let's look at the impact in practical terms. Figure 10.1 displays the predicted probabilities of civilian attacks depending on the leader's stance toward them and the structure of his group. The key takeaway is that when leaders in centralized groups oppose terrorism their members seldom commit it. Under these conditions, members attack civilians just 15 percent of the time.

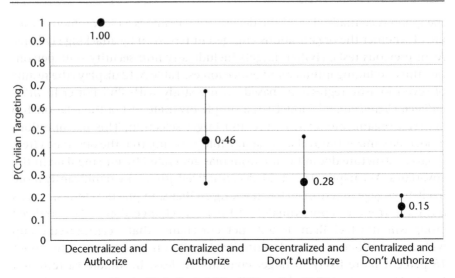

Figure 10.1. Probability of Targeting Civilians

This rate may seem high until you consider that decentralized groups always perpetrate terrorism when their leaders authorize this tactic. The most indiscriminate groups in modern times have had both of these self-destructive characteristics, including Al Qaeda in Iraq, the Armed Islamic Group, Egyptian Islamic Group, the Ku Klux Klan at its bloodiest times, and of course the not-so-brilliant Islamic State. Decentralized groups are even more likely to use terrorism than centralized groups when leaders sanction this tactic. This finding makes sense given what we know about the internal composition of militant groups. As Chapters 7 and 9 explain, the rank-and-file has stronger personal incentives to commit terrorism and is freer to pursue them in decentralized settings where the risk of punishment approaches zero. As you can see, the likelihood of subordinates flouting their targeting instructions by attacking civilians declines 13 percent when the group is centralized. This finding supports my argument in Chapter 9 that centralization helps smart leaders to communicate tactical instructions to the rank-and-file, discipline disobedient members for attacking civilians, and vet out members prone to hurting the political cause. Together, these results paint a picture of a surprisingly influential leader. The likelihood of his members using terrorism drops from 100 percent to 15 percent when he opposes this tactic and centralizes the group to make them comply. These results are robust; they hold up even after taking into account a host of alternative factors like the ideology of the group, its capability, and that of the target country. Model 3 includes the organizational covariates; Model 4 incorporates the state-level covariates; Model 5 is the same as Model 4, but with organization fixed effects; Models 6 and 7 rerun

Models 4 and 5 with imputed missing values. Across model specifications, the evidence consistently shows that centralization restrains members from engaging in terrorism, especially when the leader is known to oppose its usage.

Terrorism in Affiliates

Group structure is the most obvious measure of centralization in militant groups, but not the only one. A strong argument should hold up with different empirical strategies and data sources. Another measure of centralization is whether a militant group is an affiliate of a parent group. An affiliate is an emergent group subordinate in principle to the leadership of a more established group, viz. the parent.[2] Affiliates are a good proxy for decentralization because their members are obviously more cut-off and removed from the senior leadership in the parent group. As we've seen, Bin Laden had a more tenuous relationship with members in affiliates than in his own group. Unlike Al Qaeda Central, Al Qaeda in Iraq was a headache for Bin Laden in terms of communicating tactical restraint, punishing members for attacking the population, and screening out the worst offenders. If my theory is right that centralization promotes civilian restraint then members in affiliates should be more likely than in the parent group to commit terrorism. This relationship should be robust even after excluding Al Qaeda from the analysis and carefully controlling for organizational capability, ideology, and other factors that could plausibly influence tactical choice.

Whereas the structure of groups is debatable in some cases, parent–affiliate relationships are almost always unambiguous. The main independent variable for this analysis is whether a militant group is an affiliate of another group. With the data scientists Ryan Kennedy and Matthew Ward, I mapped the networks of militant group relationships with the Terrorist Organization Profiles (TOPs) data maintained by START.[3] These group profiles are derived from a project sponsored by the Department of Homeland Security, the Department of Justice, and the Memorial Institute for the Prevention of Terrorism. TOPs codes for twenty different types of relationships, including whether a group is an affiliate. The main dependent variable for this analysis is the attack profile of the group—specifically, the percentage of attacks carried out against civilian targets as opposed to governmental targets based on the Global Terrorism Database. The sample consists of 238 militant groups from all over the world between 1998 and 2005, as this timeframe includes the requisite data to map the global militant group networks. Figure 10.2 displays them. It plots the network graphs for all relationships between militant groups (left) and only those classified as affiliates (right). The graph for all groups is undirected, indicating any of the relationship types coded in our

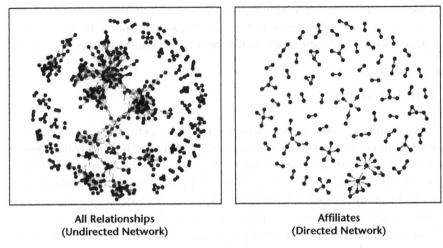

All Relationships
(Undirected Network)

Affiliates
(Directed Network)

Figure 10.2. Network Graphs of Militant Group Relationships

data. The graph of affiliates is directed, drawing a line from the affiliate to other groups. There's clearly more clustering in the all-relationships graph than in the affiliate graph. That's not surprising since relatively few militant groups have multiple affiliates.

Militant leaders evidently have a tougher time restraining members in affiliates. Table A.9 displays the results. Model 1 presents the zero-order correlation between whether a group is an affiliate and its percentage of attacks against civilians. Model 2 shows that the relationship between affiliates and civilian attacks becomes even stronger both statistically and substantively after including control variables. Once we take into account the ideology of the organizations and their capability, affiliates attack civilian targets about 13 percent more as a proportion of their total targets. Model 3 re-runs these tests without Al Qaeda, confirming that the results from Model 2 hold after removing this potential outlier. Across model specifications, the odds of this relationship occurring from chance are less than 4 percent. Figure 10.3 illustrates that compared to members in the parent group, those in affiliates are significantly more likely to direct their violence against politically suboptimal targets as leadership control wanes.

My theory also predicts that affiliates of affiliates should be even more terrorism-prone as leadership influence decreases from further decentralization. Relatively few militant groups have over two degrees of separation in the directed network as Figure 10.2 shows. Despite the limited sample size for this test, affiliates of affiliates seem to become even more indiscriminate when their network relationship to the parent becomes even weaker. Further splintering increases the percentage of civilian targets by on average about

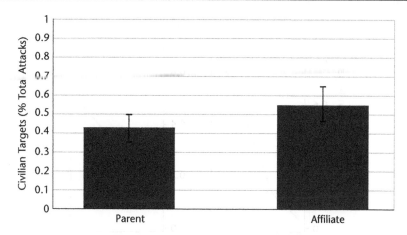

Figure 10.3. Target Selection of Affiliates Versus Parents

14 percent. Together, these results strongly suggest that members of affiliates are prone to politically risky indiscriminate violence, especially as the relationship between the senior leadership and rank-and-file becomes even more tenuous from further organizational splintering.

Terrorism in Foreign Countries

Foreign operations are another measure of decentralization. Not only are foreign theaters further away from the leader geographically, but these countries are even less sympathetic to the cause, eroding his access to operatives. When access to foot soldiers is hindered, they must rely on instruction from junior leaders or act independently, thereby weakening leadership control. Let's return briefly to the MAROB data to see whether leaders suffer a worse principal–agent problem when their members attack abroad. Conveniently, MAROB codes for whether the groups commit "cross-border" raids and "international" attacks. As anticipated, groups that commit such geographically dispersed violence are indeed significantly more likely to strike civilians. Table A.8 shows that the odds of this relationship resulting from chance is less than one in a thousand even after controlling for the capability of the groups, their ideology, and many other factors. Figure 10.4 depicts the magnitude of the effects; groups are apparently about twice as likely to target civilians in both cross-border raids and international attacks as leadership control decreases.

The results in this chapter underscore the importance of centralization for reducing politically risky civilian attacks. All else equal, militants in decentralized groups, affiliates, and operations abroad are far more likely to jeopardize

143

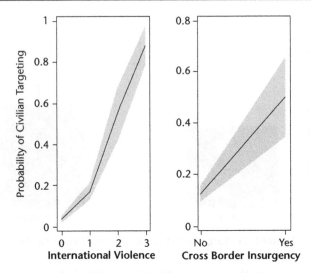

Figure 10.4. Civilian Attacks Abroad

the cause with terrorism. Under these conditions, leaders struggle to communicate tactical instructions, discipline transgressions, and screen out unreliable members. The smart leader knows not only the political costs of terrorism, but to limit its usage by centralizing the group.

Can Militant Leaders Really Centralize?

A potential objection to my argument would be if leaders lack any control over centralizing their organization. Perhaps centralization helps the leader to instill task cohesion in the ranks, but is out of his hands. Skeptics might wonder, for example, whether decentralization comes from external pressures rather than internal decisions, like a group of pigeons in a park that disperses to avoid getting trampled. Let's play devil's advocate for a moment and take seriously the counterargument that decentralization isn't a strategic choice of militant groups, but a spontaneous reaction for them to survive in the face of government repression. Admittedly, this idea has some face-value. It's not hard to think of salient historical cases when militant groups decentralized under government duress. Many Al Qaeda members hightailed it out of Afghanistan in early 2002 to escape Operation Enduring Freedom. Fifteen years later, ISIS fighters scurried from their stronghold in Raqqa, Syria, to Deir ez-Zor under a barrage of crippling U.S. airstrikes.[4] Around the same time, ISIS fighters in Libya fled from Sirte to the desert valleys and inland hills southeast of Tripoli to evade a U.S.-led onslaught.[5] Some terrorist leaders believe that decentralization is essential for organizational preservation. Louis Beam, a Klansman with strong connections to the Aryan Nations, popularized the concept of "leaderless resistance," touting the benefits of a flat, networked organizational structure to skirt government repression.[6] A common assumption among conflict scholars is that militant groups face a tradeoff between efficiency and security, so they decentralize to increase organizational survival at the expense of leadership control.[7]

But this argument is stronger theoretically than empirically. For starters, there's surprisingly thin evidence that decentralizing helps militant groups to survive. A 2008 study in *International Security* investigated this assumption and concluded that structurally diffuse, networked militant groups "are not as agile and resilient as they are made out to be."[8] The alleged benefits of decentralization rely on "only limited historical and comparative research."[9] A 2017 study in *Terrorism and Political Violence* tests hundreds of militant groups in history to determine whether decentralized ones have lasted longer. The authors find no statistical relationship between militant group structure and longevity.[10] Even more recently, Laila Wahedi finds in another large-n study with different data that militant groups in decentralized networks tend to die off faster than in centralized networks. Her detailed tests demonstrate that "Groups in centralized communities are statistically and substantively significantly less likely to end in a given month."[11] As we have seen, decentralized groups are vulnerable because they're prone to committing tactical mistakes and provoking punishing government reprisals, thereby expediting organizational demise. That's why successful leaders from Ben-Gurion to Mandela to Museveni to Nasrallah have historically punished rogue operatives. The evidence is therefore weak that decentralizing helps militant groups to survive government repression.

It's not even clear that repression is the cause of decentralization. With the data scientist Matthew Simonson, I've tested the relationship between government repression and organizational structure. It turns out that decentralization isn't an inevitable response to government repression. On the contrary, bivariate tests using MAROB data reveal no statistically significant association between government repression and decentralized groups. Once control variables like ideology are taken into account, we found that militant groups are actually significantly more likely to centralize in the face of government repression. Admittedly, research on the determinates of militant group structure remains in its infancy.[12] Yet the notion that decentralization is an automatic response to government pressure for survival simply doesn't hold water.

In fact, leaders possess considerable agency over how much to centralize the structure of their organization, its operations, recruitment, and overall decision-making. The strategic management literature in international business suggests that CEOs exercise substantial discretion over how much autonomy to delegate.[13] They have a say in how many employees to hire, what their credentials should be, and the kinds of behaviors that will get them fired. CEOs aren't passive observers about whether the firm focuses domestically or abroad. Their opinion matters about which other businesses to acquire. They don't just crowd-source strategic decisions for their company. Yes, CEOs actually do some leading. To a large extent, the decision to centralize or decentralize in militant groups also takes place at the leadership level. In the 1970s, the leaders of some left-wing groups in the Italian underground decided to experiment with decentralization, so they delegated more decisions to autonomous cells and relaxed admission standards. But these groups quickly suffered security breaches and became sitting ducks for government infiltration.[14] To survive, the Red Brigades leadership reversed course and instilled more hierarchy in the organization.[15] As I noted in Chapter 9, American militia groups underwent a similar learning process. The leaders initially chose to decentralize in order to increase their membership size. But doing so weakened operational security, allowing law enforcement to roll up the groups. Subsequently, many militia leaders opted against decentralizing for the sake of organizational preservation.

Within militant groups, decentralization has historically been internally determined. Ideology plays an important role in this decision. Take the Anarchists of the late nineteen century who valued individuality over the kind of regimentation that Marxist leaders imposed on their members. Anarchists resisted centralization not to sidestep government pressure, but to honor their convictions. Although members occasionally held conventions, they heeded Johann Most's advice: "If you want to carry out a revolutionary act, don't talk to others about it first—go ahead and do it!"[16] For radical environmentalists, decentralization is also preferred because they resist authority even in their own group. Members of Earth First! swear by their guiding philosophy of "consensus decision-making, cooperation and collective responsibility rather than representation, hierarchy and leadership."[17] Revolutionary Cells, the leftist German group active in the 1970s, also had "an extremely decentralized structure" to ensure "in our group we were all equals."[18] Similarly, decentralization is a core principle of Salafist jihadist groups.[19] Like praying, jihad is an inherently individualized act independent of centralized hierarchy. For jihadists, there is no authority other than Allah. As Nelly Lahoud explains: "Jihadis conceive of jihad as an individual duty rather than a communal endeavor organized and governed by a chain of command...Jihadism is about being driven by principles alone; organization, leadership or institutions are all deemed as burdens when the only criterion that matters is a pledge to be willing to fight and die."[20] ISIS watchers noted when operatives complied in Bangladesh, Belgium, England, France, Germany, Tunisia, and Turkey: "These attacks are not an accident. They are the result of an organized, decade-old movement within Islamic jihadism to decentralize attacks."[21] In all of these groups, the decision to decentralize came from within.

Although all Salafist groups share this belief in jihad, the leadership has tremendous influence over the extent of decentralization. Abu Musab al-Suri is technically an Al Qaeda member. But his interpretation of jihad was applied even more faithfully by ISIS leaders.[22] In his manifesto, "The Call to Global Islamic Resistance," Suri urges "solo or cellular jihad, the act of individual jihadists organizing and carrying out attacks, without any connection to or support from an established jihadist group."[23] Suri insists that such violent, decentralized acts accord with the ideals of "the jihadist movement rather than a particular organization or leader."[24] The essence of resistance, he says, is to "perform solo jihad in any part of the Arab and Muslim world, indeed in the whole world."[25] Tellingly, ISIS adopted decentralization as a guiding tenet from his playbook even before the massive international counterterrorism campaign began to shred the group.[26] By early 2014, ISIS spokesman Abu Mohammad al-Adnani was already calling on jihadists to attack independently anywhere in the world, the essence of jihad as he understood it.[27] When asked why he kills and rapes so many random people, an ISIS fighter didn't mention anything about evading government repression. "Young men need this," he told *Reuters*. And his superiors gave him "the green light."[28] When a young man in Manchester asked ISIS leaders whether he can attack some people in his country, they e-mailed him back, "OK, kill them! Show no mercy to civilians."[29] Having confirmed that ISIS heads approve of wanton slaughter, Salman Abedi bombed the Ariana Grande concert. For ISIS and other Salafist groups, the decision to decentralize targets, operations, and recruitment has clearly been based less on governmental pressure than on interpretations of jihad.

Nowhere is the role of the leader clearer than in the decision to accept affiliate groups. This crucial decision rests entirely with the parent group leader. The leader of the affiliate-wannabe pledges to him *bayat,* an oath of allegiance traditional in Islam. Affiliation is then conferred when he accepts. It's a straightforward process, except some parent leaders are more discriminating than others about accepting the pledge. Just ask Boko Haram. Its leader, Abubakar Shekau, courted Al Qaeda in July 2010. But bin Laden was smart enough to reject the offer because he knew the Nigerian group's penchant for civilian massacres would further besmirch the Al Qaeda name.[30] As one analyst noted: "The attacks on civilians and particularly these girls is [sic] not something he would sanction. Boko Haram has crossed the line."[31] Fast-forward to March 2015 and Boko Haram had a very different outcome—not with Al Qaeda, but Islamic State. Shekau pledged Boko Haram's allegiance to the Islamic State in an audio recording that followed a script used by many other affiliates of the Islamic State. A few days later, an ISIS spokesman announced

Baghdadi's acceptance of Shekau's pledge. The courtship worked. Other ISIS affiliates then released celebratory videos welcoming the civilian-ravaging "Nigerian mujahedeen."[32] Unlike Bin Laden, Baghdadi had no compunction about their bloodlust. This was, after all, ISIS.[33] By 2016, he had accepted pledges from forty-three affiliates all over the world, including ones Zawahiri had recently rejected as too extreme for Al Qaeda.[34]

Terrorism analysts in the media have extolled the decentralization of ISIS as highly strategic. Clint Watts wrote in *War on the Rocks*: "The Islamic State accrues straightforward benefits accepting affiliates."[35] In *Foreign Affairs*, he characterized the ISIS open-door policy as a "spectacular" strategic decision.[36] Will McCants claimed the indiscriminate absorption of the most unhinged people and affiliates around the world was "beneficial" to ISIS.[37] Malcom Nance exclaimed, "The brilliance of the ISIS system is that its recruitment system is almost passive..."[38] Peter Bergen agrees that ISIS is "victorious" largely because it "accepts all comers."[39] Baghdadi was unfazed about giving up control in communicating to these new members, disciplining them, or vetting out bad ones. These levers of restraint—indeed the utility of restraint itself—never seem to have crossed his mind.[40] For better or worse, militant leaders clearly have a big say over the structure of their group, its adoption of affiliates, and the decision to authorize attacks abroad. In the short term, such decentralization helps groups to maximize carnage. But that's the opposite of what successful rebels do.

Conclusion

With few exceptions, the trend in academic and policy circles is to tout decentralization as an unconditional "best practice" for militant groups.[41] Research into their structure has hailed the benefits of decentralization since the publication of *Networks and Netwars* in 2001.[42] In their seminal work, the political scientists, John Arquilla and David Ronfeldt, spell out the advantages of a diffuse organizational structure. As with other organizations, decentralization is thought to make militant groups more adaptive, flexible, inclusive, innovative, and specialized. In terms of defense, decentralization supposedly makes militants harder to detect, infiltrate, isolate, prosecute, and ultimately defeat. In its purest form, the structure is a flat network rather than a hierarchical organization. There is no central command. All decision-making is delegated to autonomous members of the rank-and-file. In fact, everyone is rank-and-file. The smartest leaders are ones who don't exist at all. The *9/11 Commission Report* affirms that terrorist networks are more "agile, quick, and elusive" than centralized terrorist organizations.[43] *The Economist* reports that decentralizing enables terrorist groups to be more "fluid, mobile, and incredibly resilient."[44] Yet none of these analyses shows that decentralizing contributes to political success or longevity. Ironically, the militant groups identified by Arquilla and Ronfeldt as the most decentralized have been the least successful according to their own members, from the Algerian Armed Islamic Group to the Egyptian Islamic Group.[45]

The application of principal–agent theory predicts inherent tradeoffs to delegation.[46] As we saw in Chapter 7, the modal militant leader presides over fighters who are indifferent to the political cause, clueless about how to achieve it, and incentivized to subvert it with terrorism. Should the smart rebel leave his political fortunes in the hands of such clowns? No, he doesn't just throw up his hands and tell every schmuck in the world to wage their own jihad. Rather, he centralizes the group to harness the power of its members for the cause. In Chapter 11, we'll examine what happens to the tactical choices of militant groups after a decapitation strike when they're suddenly left with no leader at all.

Notes

1. This analysis is based on a co-authored paper on a related, but different topic. See Abrahms and Potter 2015. The paper is related because we assessed whether decentralized militant groups are prone to principal–agent problems. The paper has a fundamentally different argument, however; it doesn't take a stance on whether leaders should centralize their group. Centralization is treated as a proxy of principal control rather than a smart strategic choice from the top.
2. Byman 2014.
3. Abrahms, Ward, Kennedy 2018.
4. "Islamic State Bureaucrats Fleeing," February 17, 2017.
5. Lewis, February 10, 2017.
6. Beam 1992, 162–3.
7. See, for example, Drake 1998, 164; Cronin 2011, 102; Laqueur 2016, 99.
8. Eilstrup-Sangiovanni and Jones 2008, 17.
9. Ibid. 9.
10. Pearson et al. 2017.
11. Wahedi 2018.
12. Staniland 2014, 3; Kilberg 2012, 812.
13. Mintzberg 1989; Eisenhardt 1989; Davis et al. 1997; Dau, Moore, Abrahms 2018.
14. Della Porta 1995.
15. Snow 1999, 131; Eilstrup-Sangiovanni and Jones 2008, 30.
16. Boot 2013, 232.
17. Purkis 2001, 161.
18. Della Porta 2006, 118.
19. Maher 2016, 40.
20. Lahoud 2010, 14, 144.

21. Bazzi, July 26 2016.
22. Rej 2016.
23. "Abu Musab al-Suri's Military Theory," 2017.
24. Ibid.
25. Ibid.
26. Rej 2016.
27. Clarke and Gartenstein-Ross, November 10, 2016.
28. Georgy, February 17, 2017.
29. Byrne, August 14, 2017.
30. Bipartisan Policy Center 2014.
31. Cocks, May 28, 2014.
32. Almukhtar, June 11, 2015.
33. In August 2016, Boko Haram named a new leader, Abu Musab al-Barnawi, to replace Shekau.
34. McCants 2015, 141.
35. Watts, June 13, 2016.
36. Ibid, April 4, 2016.
37. McCants 2015, 141.
38. Nance 2016, 210–11.
39. Bergen, November 5, 2015.
40. Intel Center 2015.
41. Paul 2007, 352; Eilstrup-Sangiovanni and Jones 2008. Exceptions include Cunningham 2013; and Mahoney 2017.
42. Arquilla and Ronfeldt 2001.
43. Kean and Hamilton 2004, 399.
44. "Waiting for al-Qaeda's Next Bomb," May 3, 2007.
45. Arquilla and Ronfeldt 2001, 32–3.
46. Hawkins et al. 2006.

11

When Elephants Rampage

Some years ago, the Kruger National Park in South Africa faced an elephant problem. The population of African elephants had grown too large for the park to sustain. So, measures were taken to thin the ranks. The plan was to relocate the elephants to another reserve. A harness was constructed to air-lift the elephants by helicopter to the Pilanesburg National Park. But the harness could only handle the juvenile and adult female elephants. A decision was reached to leave the bigger African bull elephants at Kruger. At first, this solution seemed to work; everything was normal at Kruger. Yet rangers in the Pilanesburg park noticed a strange problem at the elephants' new home—dead bodies of endangered white rhinoceroses. The rangers were confused about the culprits. Surely, poachers would have taken the precious tusks. And the rhinos died from deep puncture wounds, not gunshots. Hidden cameras were set up throughout the park to get to the bottom of this mystery. The rhino killers turned out to be marauding bands of aggressive juvenile male elephants, the very animals that had just been air-lifted from Kruger. Caught on video were the young males knocking down the rhinos, stomping them, and goring them to death with their own tusks. The park rangers settled on a hypothesis to explain this bizarre behavior. Perhaps the relocated herd was acting up without the senior bulls to accompany them. In the wild, adult bulls were known to act as models for younger elephants to emulate, keeping them in check. Missing their elders to restrain them, the younger elephants went berserk, lashing out in indiscriminate fits of violent rage. To test this explanation, the rangers constructed a stronger harness and flew in the most senior bulls from Kruger. Almost immediately, the younger elephants fell into line and the rampages stopped.[1]

This chapter examines the effects of removing not bull elephants from herds, but leaders from militant groups in so-called decapitation strikes. Decapitation isn't a new strategy. The Sicarii employed assassinations in Jerusalem as early as 60 AD. Israel has practiced targeted killings ever since Palestinian terrorists from Black September murdered eleven members of its

Olympic team in the 1972 Munich Games.[2] The U.S. embraced targeted killings after the 1998 East Africa embassy bombings, when President Clinton issued three top-secret Memoranda of Understanding to kill Bin Laden and his lieutenants.[3] In practice, though, targeted killings remained relatively rare until Al Qaeda struck the homeland. Armed with rapidly improving drone technology, President George W. Bush authorized nine targeted killing attempts from 2001 to 2007, followed by thirty-six in his final year of office. President Obama dramatically intensified drone strikes against militants in the tribal regions of Pakistan, as well as in Iraq, Libya, Somalia, Syria, and Yemen.[4] The Trump administration relaxed targeting guidelines even further, ramping up drone strikes to unprecedented levels.[5]

Most research on leadership decapitation assesses whether it "works" in reducing the lifespan of militant groups or their ability to produce violence.[6] But my theory suggests a more nuanced finding about the effect of decapitation strikes on tactical decision-making. If militant leaders often restrain the rank-and-file, then taking them out should make their groups even more extreme in their targeting choices. Without the leader communicating which targets to avoid, punishing transgressors, and vetting out rogue operatives, they're freer to act on their own initiative to attack civilians. More than the quantity of violence, decapitation reduces its quality. This chapter presents a battery of evidence that removing the leaders in most militant groups leads to more indiscriminate violence by empowering lower-level members who exercise less civilian restraint. The test of a leader is whether a group is more effective with him at the helm.[7] The telltale sign of a really stupid leader is when taking him out has no negative impact on the quality of the group's tactics.

Anecdotal Evidence

Anecdotal evidence suggests that removing the leader of a group may lead it to rampage like unbridled elephant herds. In January 2016, Mexican marines captured Joaquín "El Chapo" Guzmán, the longtime head of the Sinaloa cartel. Rather than reducing the gang violence, taking him off the streets made them bloodier than ever. Not only did the amount of violence spike, but so did the target selection to include innocent bystanders. A gang member compared the type of cartel violence before and after the arrest: "If we wanted to kill you and you turned up with your wife and children, we couldn't do anything. We couldn't touch you. Now, they don't give a damn . . . If they see you in a taco stand, they'll come and shoot it up."[8] Criminologists have found an uptick in community violence when gang leaders are removed. In his study on "the adverse effects of arresting a gang's leader," Robert Vargas noted a rise in violent crime after imprisoning Rudy Cantu, the 22 Boys leader. Left to their own devices, the gangsters went on a shooting spree throughout Chicago.[9] The applied economists, Jason Lindo and María Padilla-Romo, measured the effects of targeting high-ranking gang members on Mexico homicide rates between 2001 and 2010. This "kingpin strategy" increased homicides by 80 percent in the municipalities where the leaders had operated.[10] Nonviolent protesters have also escalated when their leaders were arrested. In July 1964, the Congress of Racial Equality (C.O.R.E.) held a rally to protest a shooting in Harlem by an off-duty policeman. C.O.R.E. organizers led about a hundred people to the precinct station demanding an end to police brutality. The organizers threatened to sit in the street until the demand was met. Scuffles broke out when police pushed the crowd back, but the organizers maintained control. This control lapsed, though, the moment they were arrested. With the leadership gone, the peaceful sit-in erupted into an uncontrollable mob throwing bricks and bottles.[11]

Historically, militant groups have also become less restrained upon losing their leaders. In 1954, the British launched "Operation Anvil" to stamp out the Mau Mau uprising. Capturing their leaders around Nairobi initiated a period of uncoordinated, rudderless violence.[12] ANC tactics likewise became less disciplined when its leadership was sidelined. In 1961, the ANC established an armed wing called Umkhonto we Sizwe, which came to be known as the MK. As we've seen, the leadership stressed the use of "properly controlled violence" to spare civilians.[13] For three years, MK members complied by studiously avoiding terrorist attacks. Then, at the 1964 Rivonia trial, Mandela was sentenced to life imprisonment. ANC restraint ebbed during his years in confinement. As the decade unfolded, young men engaged in stone throwing, arson, looting, and brutal killings of civilians.[14] Gregory Houston observes that "the removal of experienced and respected leaders...created a leadership vacuum" that empowered hot-heads out for vengeance.[15] PIRA violence also became less discriminate after the leaders were neutralized, as a study recounts: "When much of the PIRA leadership was interned in 1971–2, one result was that young, aggressive, and undisciplined terrorists perpetrated a greater level of uncoordinated and indiscriminate violence than seen before."[16] When Filipino police assassinated its founder, Abdurajak Janjalani, in 1998, the Abu Sayyaf group devolved into a movement of bandits that preyed on private citizens.[17] When Nigerian police summarily executed its founder, Mohammed Yusuf, in 2009, Boko Haram became more ruthless towards civilians, making "the good old days of Yusuf's reign seem benign."[18] The Salafist rebel group Ahrar al-Sham also became more extreme after a 2014 attack on its headquarters in the northwestern province of Idlib took out the leadership.[19]

Removing leaders in gangs, civil resistance movements, and militant groups has eroded tactical restraint by empowering lower-level members. My research suggests that leadership deficits in militant groups should increase terrorist attacks against civilians, as subordinates with weaker civilian restraint gain tactical autonomy without the leader communicating to them which targets to engage, punishing violators, and vetting out terrorism-prone operatives. Let's inspect a couple of qualitative case studies more carefully to unpack the internal dynamics of militant groups when their leaders are taken out. Then, we'll quantify the effects of targeted killings on militant group tactics.

The Case of the Al-Aqsa Martyrs' Brigade

Targeted killings during the Second Intifada created leadership deficits in the Al-Aqsa Martyrs' Brigade that spurred terrorist attacks against Israelis. The historical record reveals that (1) the Brigade leadership wanted to avoid civilian targets in order to create a Palestinian state; (2) Brigade members turned their violence against Israeli civilians as the Intifada unfolded; (3) the diminished targeting selectivity was due to a loss of leadership control from Israeli decapitation strikes; (4) whereas the leadership recognized that attacks on civilians are politically counterproductive, lower-level members perpetrated them for personal reasons based on their position within the organizational hierarchy.

As the military wing of Yasser Arafat's secular Fatah Party, the Brigade was established in September 2000 to pressure Israel into withdrawing from territories captured in the 1967 war. To end the occupation of Jerusalem, West Bank, and Gaza Strip, Brigade chief Marwan Barghouti called for selective attacks against the Israel Defense Forces (IDF) and settlement outposts, but to spare Israeli civilians within the pre-1967 borders or so-called Green Line. Barghouti stated in interviews: "We said we would not attack inside the Green Line. The real face of the occupation is the settlements and the soldiers."[20] He emphasized that "Fatah's line is only targets outside of [19]67 borders,"[21] that "Our policy in Fatah has been to restrict our actions to the territories,"[22] and that "I, and the Fatah movement to which I belong, strongly oppose attacks and the targeting of civilians inside Israel, our future neighbor."[23] Fatah leader, Hussam Khader, stressed the strategic benefit of selective violence against Israelis, "When they realize that there are no civilian casualties and only soldiers dying in a foreign land, it will spark a change we need on the Israeli street to bring an end to the occupation."[24] Even Arafat expressed "total opposition to actions targeting civilians on both sides."[25] This position went largely unchallenged among lower-level Brigade leaders. The head of the Bethlehem unit during the Second Intifada declared that harming Israeli civilians is "completely unacceptable to us in al-Aqsa" and that instead "Our strategy is to fight settlement and settlers [by] attacking Israeli military posts."[26] A Ramallah-based Brigade leader reiterated that "I am against touching civilians," though strongly supportive of hitting the IDF and other instruments of the occupation.[27]

Initially, Brigade members complied with these targeting guidelines. In late 2001, operatives attacked the IDF in Haifa, Hebron, and Tel Aviv. By early 2002, however, the Brigade committed mass-casualty attacks at a bar-mitzvah in Hadera, a kibbutz in Menashe, and the Tel Aviv Central Bus Station. The reduced targeting selectivity didn't go unnoticed. The National Consortium for the Study of Terrorism and Responses to Terrorism (START) observed: "At the outset, al-Aqsa Martyrs Brigade expressly targeted Israeli settlers and secur-ity forces. However, the group soon expanded its targets to include citizens in Israeli cities."[28] The Council on Foreign Relations also remarked: "While the group initially vowed to target only Israeli soldiers and settlers in the West Bank and Gaza Strip, in early 2002 it joined... in a spree of terrorist attacks against civilians in Israeli cities."[29] The U.S. State Department listed the Brigade as a terrorist organization during this unprecedented wave of indis-criminate bloodshed.

It stemmed from a loss of leadership control. The locus of decision-making became disjointed in early 2002, as the IDF killed off dozens of Brigade commanders, culminating with the arrest of Barghouti that spring.[30] According to *The Economist*, "Fatah's resistance went from guerrilla warfare to freelance martyrdom operations inside Israel" because of "the increasing autonomy of the militias" and "widening gulf between political and military wings."[31] The political scientist, Wendy Pearlman, notes that leadership deficits increased because Palestinian leaders were not only taken out, but forced into hiding, thereby severing "communication and coordination" with subordinates.[32] The International Crisis Group reported that as a result of the decapitation strikes, "The network is diffuse, fragmented, localized, and does not take orders from leaders of the organization."[33] Human Rights Watch also described the centrifugal effects of the decapitation campaign: "The military elements responsible for the [terrorist] attacks are not under the control of the political leadership" because "there is no infrastructure, just small groups making their own small decisions" with "a [large] degree of autonomy and improvisation."[34] A Palestinian Authority minister lamented that the removal of the Brigade leadership left "different factions and fractions within factions pursuing different strategies."[35] An Arab-Israeli journalist underscored the growing disconnect between principal preferences and agent actions: "Most of the military operations are being carried out by gunmen who don't report to their political leaders. Even if the factions had reached an agreement, this wouldn't have meant a complete end to the [terrorist] violence."[36] A Palestinian intellectual affirmed the newfound independence of operatives, "The decision to resist was taken independently in the [Jenin] camp, in violation of the leadership's orders."[37] When asked about the Brigade's target selection, even the militants acknowledged that "not all military acts by al-Aqsa were done with the agreement of the political wing" as "professed identity with Fatah did not necessarily translate into compliance with Fatah decisions."[38] To explain the spasm of violence, a refrain heard throughout the Palestinian territories was "There is no leadership."[39] With the Brigade leadership sidelined, the young men rampaged.

Although Brigade leaders knew the indiscriminate violence was politically counterproductive, its operatives were driven by alternative incentives based on their position within the organization. Barghouti opposed Palestinian attacks on Israeli civilians because he had observed over the years how the "impact on Israeli public opinion was detrimental to us."[40] Arafat too had come to learn the strategic perils of civilian targeting, warning in the largest Palestinian daily: "Actions that target civilians are counter to the lofty interests of our nation, hurt the legality of its legitimate struggle against the occupation, and cause damage to its image."[41] Other Fatah leaders also seemed to appreciate the costs of attacking civilians. Khader, for example, lamented that "they unite the world against us"[42] and al-Sheikh worried that "they have reduced the level of international support for the Palestinian people."[43] In May 2002, the 130-member Fatah Revolutionary Council issued a statement condemning "military operations inside Israel . . . because they are likely to have a negative impact on national resistance."[44] The following month, dozens of Palestinian leaders released an even stronger statement in *al Quds*: "We call upon the parties behind military operations targeting civilians in Israel to reconsider their policies . . . these bombings do not contribute towards achieving our national project . . . On the contrary, they strengthen the enemies of peace on the Israeli side."[45] But the senior leadership wasn't representative of the other members.

Those who filled the leadership void tended to be younger, less experienced, disciplined, and strategic.[46] The rank-and-file were also less political and committed terrorism for personal reasons. Operatives reportedly perpetrated terrorism so the local community would "look up" to them and as a "power grab" to advance within the organization by outbidding more restrained rivals.[47] The International Crisis Group emphasized how "above all" lower-level members were motivated by "struggles for power and position" within the Brigade.[48] Other observers pointed to the role of blood revenge among foot soldiers at the front-line. The *New York Times* reported that Brigade operatives committed terrorism "often in revenge for Israeli killings" of their loved ones.[49] A piece in the *New York Review of Books* also noted: "In many cases the bombers say they are taking revenge for the death of someone quite close to them, a member of their family or a friend."[50] A demographic study on Palestinian operatives during the Second Intifada found that "revenge was their primary motive" unlike that of the leadership.[51]

The Al-Aqsa Martyrs' Brigade provides insight into the decision-making process of a militant group when low-level participants are left in charge. As Yezid Sayigh of the Carnegie Middle East Center remarks, "Internal dynamics help explain the often chaotic and counterproductive nature of Palestinian military activity."[52] This was manifestly the case during the Second Intifada, when Brigade violence became less selective due to a loss of leadership control. Although the leadership understood the political costs of attacking the Israeli populace, the decapitation campaign empowered lower-level members bent on attracting esteem within the community, climbing the organizational hierarchy, and avenging Palestinian suffering to the detriment of the cause.

The Case of the Afghan Taliban

Targeted killings of the Taliban also had the perverse effect of creating leadership deficits that promoted indiscriminate violence because (1) the upper echelon had commanded lower-level members to attack military targets while punishing civilian attacks; (2) subordinates in the Taliban tend to be younger, less experienced, and strategic in their thinking; (3) targeted killings have changed the internal composition of the organization; and (4) the erosion of control from the top has empowered these fighters with inferior civilian restraint.

Since it was overthrown in the post-9/11 American-led invasion, the Taliban has killed thousands of Afghan civilians mainly with suicide attacks, improvised explosive devices, and assassinations. These actions are in direct defiance of the leadership's injunctions to safeguard the population. According to the Taliban leadership, its fighters are prohibited from attacking civilians because such indiscriminate violence risks "ending up in a clash between civilian and Taliban."[53] Mullah Omar, the founder and chief strategist of the Taliban until his mysterious death in April 2013, emphasized in public statements, "The mujahedeen have to take every step to protect the lives and wealth of ordinary people."[54] This position cannot be dismissed as merely propaganda; internal documents seized by U.S. military forces confirm that the upper echelon of the Taliban regards civilian attacks as a political liability.[55]

The leadership's proscription against harming civilians has been at the core of the code of conduct since the first Layha of 2006. Rule 21 states, "Anyone who has killed civilians during the Jihad may not be accepted into the Taliban movement."[56] Rule 46 declares, "Taliban commanders should try their best to avoid civilian deaths during fighting."[57] Rule 41 in the 2009 Layha reminds operatives to "avoid civilian casualties"; rule 48 bans "cutting noses, lips, and ears off people"; and rule 59 mandates, "The Mujahidin must have a good relationship with all the tribal community and with the local people."[58] Rule 57 in the 2010 Layha decrees, "In carrying out martyrdom operations, take great efforts to avoid casualties among the common people." Rule 65 enjoins mujahedeen to "be careful with regard to the lives of the common people and their property"; and the back cover stresses that "taking care of public property and the lives and property of the people is considered one of the main responsibilities of a mujahed."[59] The Taliban Leadership Council emphasizes that the only permissible targets are selective: " . . . foreign invaders, their advisors, their contractors and members of all associated military, intelligence and auxiliary departments. And similarly, the high-ranking officials of the stooge Kabul regime; members of Parliament; those associated with Ministries of Defense, Intelligence and Interior."[60] These targeting guidelines are also enshrined in the 2010 code of conduct. Rule 5 directs operatives to attack "high-ranking government officials" and rule 41 demands the target of suicide bombers to be "high valued."[61] Since then, the leadership has only increased its warnings to operatives about the political perils of harming the population.[62]

Beyond targeting instructions, the leadership has tried to incentivize fighters to refrain from counterproductive civilian attacks. The Code of Conduct isn't devoid of enforcement mechanisms.[63] According to the NATO-led International Security Assistance Force, the injunctions against civilian harm are "strictly enforced, with an elaborate system of checks and balances to insure [sic] compliance." To gain local support, the Taliban leadership has established independent commissions throughout Afghanistan of "neutral observers and judges" to formally complain about targeting violations.[64] Taliban leaders have also distributed phone numbers for anonymous complaints, leading to numerous cases in which offending members were expelled from the Taliban, stripped of their position, disarmed, imprisoned, or publicly rebuked. When found guilty of civilian targeting, commanders have been sent to Mullah Omar and other high-level leaders for punishment under Sharia law.[65] Not only are targeting violations subject to punishment, but Taliban leaders educate foot-soldiers about best practices, offer hands-on training to spare civilians, steer them away from crossfire, and promote members for engaging military targets.[66] A United Nations report concludes that Taliban leaders thus reduce "casualties among the common people" by "implementing guidance in the Layha to target military objects more carefully."[67] Despite these efforts, Taliban foot-soldiers have killed thousands of Afghan civilians.

These attacks are often the result of a principal–agent problem. The Afghanistan-based journalist Kate Clark notes, "The Taliban has severe command and control problems within its ranks."[68] The International Crisis Group observes that the leadership "has struggled to exert authority over its field commanders" and that "Given the autonomy that Taliban commanders and allied networks enjoy, the leadership might exercise little control over every-day military operations."[69] The U.N. Secretary General for Afghanistan agreed that actions "by the Taliban leadership are not nearly enough to end the killing and injuring of innocent Afghan civilians."[70] *Reuters* adds, "Even if the Taliban wants to bring down the number of civilians it kills, it lacks total control over the bombers, or those who guide them."[71] The *New York Times* affirms that attacks against civilians are typically perpetrated by low-level members acting "on their initiative" because "The Taliban is a fractious organization and its leadership ... often has only marginal control over the day-to-day activities of the rank and file, who usually decide whom to attack and when."[72] Gen. John R. Allen, who commanded the American-led coalition, likewise remarks that low-level members have been "the ones who were planning the roadside bombs and intentionally targeting civilian targets" seemingly "isolated from more senior Taliban leadership."[73]

Since 2008, the U.S. has contributed to this principal–agent problem with an intense targeted killing campaign. On the surface, the impact of the drones has been modest. The Taliban's ability to mount operations has stayed constant, as a seemingly endless supply of lower-level members has ascended through the ranks. The Afghanistan specialist, Anand Gopal, observes that the removal of over a thousand "high-value" Taliban members "has not diminished the insurgents' capability to attack ... What it has done is force major demographic shifts in the makeup of the insurgency and a concomitant shift in insurgent operating procedures."[74] A report from the Center on International Cooperation at New York University remarks: "The military campaign targeting insurgent leaders within Afghanistan has weakened the overall command structure and the ability of the central leadership to enforce decisions. A reshuffling of the leadership, along with all layers of ranks of commanders, has seen the rise of a younger and more radical generation."[75] Given the internal dynamics of the Taliban, the targeted killings have eroded organizational restraint towards civilians by endowing subordinates with additional operational autonomy. According to an estimate from the Human Terrain System, targeted killings have reduced the average age of Taliban leaders from 34 to 26 years old.[76] This estimate is consistent with Al Qaeda's impression. Before being droned in Waziristan in 2011, Atiyah Abd al-Rahman lamented that targeted killings result in "the rise of lower leaders who are not as experienced as the former leaders."[77] Indeed, there's broad agreement that whereas the older guard helped to impose selective violence, replacements are more inclined to harm the population because they're "less experienced and less skilled,"[78] "less competent,"[79] "prone to make operational and strategic mistakes,"[80] "less likely to be amenable to restraining their actions" against civilians,[81] "more brutal" towards civilians,[82] and "more radical"[83] in their treatment of the population.

As with the Al-Aqsa Martyrs' Brigade, the Taliban case provides insight into the importance of leaders for restraining lower-level members from harming civilians. When their leaders were taken out or forced into hiding, poorly disciplined members gained autonomy and rampaged against the population. These qualitative cases suggest that leadership deficits lead to more terrorism by unshackling lower-level members of militant groups with weaker civilian restraint. The quantitative tests below substantiate this impression.

The Impact of Decapitation Strikes on Tactics

The New America Foundation's Drone Database contains fine-grained data on the ongoing drone campaign in the Afghanistan–Pakistan (AfPak) tribal region. The Drone Database supplies information on the timing of decapitation strikes and operational outcomes—both of which should affect leadership influence over the rank-and-file. When a drone strike kills a leader, his subordinates are given a freer hand in conducting operations. Even when drone strikes don't connect with the target, they force him and other leaders to assume a diminished posture within the organization for their survival, strengthening the influence of militants. In an earlier study, Phil Potter and I paired these drone strike data from the New America Foundation with information on the militant groups' target selection from the Global Terrorism Database.[84] Given what we know about the internal dynamics of militant groups, one can assess whether they're more likely to direct their violence against civilian targets after a drone strike has reduced the influence of the leader by killing or forcing him into hiding.

Table A.13 displays the results. They provide additional evidence that leadership deficits from targeted killings result in more indiscriminate violence. When a drone strike kills a leader, the group increases attacks on civilians while decreasing attacks on military targets. The effects of targeted killings are strong both statistically and substantively. For instance, predicted civilian targeting rises by about 40 percent after just one leader gets killed. And regardless of the target's fate, drone strike density also promotes indiscriminate violence as leaders take cover. To illustrate the change in group tactics, the left panel of Figure 11.1 plots the predicted number of civilian attacks based on the rate of drone strikes in that theater. As you can see, the number of militant group attacks per day on civilians nearly triples depending on the rate of drone strikes. The right panel reveals how much killing the leader lowers the likelihood of attacks against military targets. We see a reduction in the rate of attacks against hard targets from one every five days to one every twenty days when two leaders have been successfully neutralized in the prior week.[85]

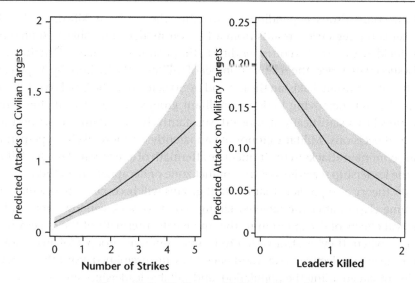

Figure 11.1. Civilian and Military Targeting

These preliminary results indicate that militant groups in the Afghanistan–Pakistan tribal region are far more likely to direct their violence against civilians as the leader's control recedes by his death or by forcing him into hiding to escape a barrage of drone strikes around him. Although both drone killings and misses reduce leadership control, we would expect the former to have an even larger effect than the latter on militant group tactics because leaders can still exert some influence while in hiding. Even when on the run, for example, Bin Laden continued to supply tactical instructions and reprimand wayward fighters remotely.[86] If my take on militant group tactics holds, we should therefore find that operationally successful drone strike attempts have an even bigger impact than misses on the proclivity of militants to attack civilians. We should also find that successful targeted killings promote indiscriminate violence in other conflict zones too, as well as when leaders are neutralized with lethal instruments other than drones, like a bullet to the head. A dataset from Asaf Zussman and Noam Zussman facilitates this assessment by including information from the Israeli targeted killing campaign against the most lethal Palestinian groups in the West Bank and Gaza Strip during the Second Intifada. Unlike the New America Foundation data, the Zussman and Zussman dataset contains decapitation methods beyond drones, such as from snipers and bombs.[87]

The tactical change in the organizations is remarkably consistent across conflict zones. In both the Afghanistan–Pakistan and Israel–West-Bank–Gaza theaters, the ratio of attacks on military targets versus civilian ones is markedly lower in the two-week period after an operationally successful strike, when tactical decision-making is presumably most affected. Table A.14 presents the mean comparisons depending on whether the leader is killed; Table A.15 presents the regression results from a Poisson model; and Table A.16 presents the incidence rate ratios corresponding to the parameter estimates.[88] In the wake of a successful strike, the relative number of military attacks drops by 30 percent in the Afghanistan–Pakistan theater and by 50 percent in the Israel–West-Bank– Gaza theater. Removing leaders of militant groups in these two highly active theaters has a statistically significant, substantively important impact on their tactical decisions. Militant groups are apparently far more likely to perpetrate indiscriminate attacks in the immediate aftermath of a successful targeted killing of the leadership regardless of the conflict zone or decapitation method.

My theory also predicts that militant groups will become more discriminate as time elapses after the targeted killing. The two-week window captures the tactical effects of decapitation in the immediate aftermath of the government strike, when they're strongest. Over time, militant group violence should become more selective as wayward operatives are disciplined, learn the strategic value of safeguarding the population, and suitable leadership replacements rise to the top. As an additional test, I therefore extend the window of one to two weeks to three to four weeks and five to six weeks after the targeted killing attempt. The results presented in Table A.17 confirm that militant group tactics indeed become less indiscriminate over time. Throughout these analyses, I've also paid attention to time in other ways, too, to better assess the impact of leadership deficits on civilian targeting. Seasonal effects are potentially important because civilians are more likely to spend time outdoors and thus be attacked as weather conditions improve. Drone operators say that clouds and precipitation can obscure satellites, hindering operations. Declassified Al Qaeda documents reveal that Bin Laden warned operatives against switching safe houses on clear days.[89] To alleviate such weather-related concerns, I include a specification with monthly dummies. I also control for Ramadan, a religious holiday that may also affect the exposure of Muslim populations to militant attacks as well as government countermeasures. For neither theater are the findings driven by the periodization scheme, seasonal variation, or Ramadan festivities. Taken together, these tests provide strong evidence that militant groups gravitate to terrorism when they're suffering from leadership deficits and lower-level members are calling the shots. Militant groups are more likely to direct their violence against civilians in the immediate aftermath of decapitation strikes, especially when they register a kill.

Not All Militant Leaders Are the Same

Of course, not all leaders matter equally in militant groups. Decapitation strikes have different effects depending on the role of the leader. The Zussman and Zussman dataset distinguishes between political leaders, who mainly supply ideological guidance, and military leaders, who mainly supply tactical guidance. There's no significant change in the target selection of militant groups when their political leader has been killed. By contrast, the violence becomes even more indiscriminate both statistically and substantively when the sample of decapitated leaders is restricted to military ones. Clearly, militant groups aren't simply lashing out against civilians to avenge the loss of a leader. A reduction in the quality of organizational violence happens only when tactical guidance is compromised from degrading military command.

If I'm right, we should see a larger tactical change when the most senior militant leaders are killed. Differences in the targeted killing datasets enable this assessment. The Zussman and Zussman dataset is comprised of targeted killing attempts against only the most senior leaders, whereas the New America Foundation dataset also includes attempts against mid-level leaders and even low-level militants. Because of my interest in the tactical effects of decapitation, I have consistently excluded from the analysis attempts against non-leaders. Yet exploiting the variation in militant leader-type can help to evaluate whether taking out more senior leaders has a greater impact on civilian targeting. Table A.16 confirms that the relative number of military attacks after a targeted killing is 20 percent lower in the Israel–West-Bank–Gaza theater than in the AfPak theater presumably because the decapitated Palestinian leaders had occupied more senior positions within their organizations. Militant groups should also experience larger tactical shifts when the leader was geographically closer to operatives and therefore better positioned to influence their targeting choices. The New America Foundation codes for whether the decapitation strikes happened in Pakistan or Afghanistan. As anticipated, organizational violence becomes even less discriminate in Pakistan following a successful targeted killing where most of the AfPak strikes have been carried out.[90]

Throughout Part 2 of this book, we've seen that militant leaders restrain members to varying degrees from harming civilians. That's why most militant groups become more indiscriminate in their target selection after losing a leader. The effects are obviously greatest in groups whose leaders opposed terrorism because replacements aren't expected to share their targeting preferences. Decapitation can also make a group less discriminate even when it was led by a strong leader who supported terrorism because then members will have an even freer hand to act on their own initiative against civilians. As we saw in Chapter 10, groups led by weak leaders who support terrorism are the only ones who invariably commit it. You can see in the 2-by-2 chart depicted in Figure 11.2 that only one type of group won't attack more civilians when the leader is killed—groups led by weak leaders who favored terrorism. This type of group is already so unrestrained in its targeting choices that taking out

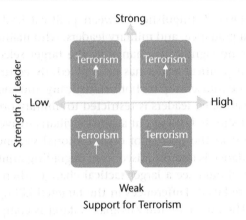

Figure 11.2. Effect of Decapitation Strike on Terrorism

the leader can't make members any more indiscriminate. Such leaders neither understood the dangers of indiscriminate violence nor built the organization to limit it by communicating which targets to avoid, punishing transgressions, or vetting out worthless members. The leader is so incompetent that his death has no impact whatsoever on their performance.

An example is Baghdadi. In May 2017, the Russian Defense Ministry claimed in a Facebook post to have killed the ISIS leader in an airstrike outside Raqqa. Excited reporters seemed crestfallen when I told them the story was likely apocryphal and didn't matter anyway. With its modus operandi of decentralized carnage, ISIS fighters will rampage with or without their stupid leader. The loss of Baghdadi is irrelevant to the tactical choices of other members because he failed to recognize the political costs of civilian attacks or to design the organization to restrain them. The first two parts of this book have shown that Baghdadi violated both of those rules for rebels. In the next section, we'll see that he violated the third and final rule as well. Smart leaders not only minimize terrorism, but distance themselves from it to brand the organization as more moderate than its actual behavior.

Notes

1. Macrae, June 20, 2012.
2. Byman 2011.
3. Zenko 2012, 62–3.
4. Kaag and Kreps 2014.
5. Telesur, June 13, 2017.
6. See, for example, Johnston 2012; Jordan 2009; Price 2012.
7. Nye 2008, 20.

8. Agren and Juarez, May 5, 2017.
9. Vargas 2014, 143.
10. Lindo and Padilla-Romo 2015.
11. Sharp and Finkelstein 1973, 617.
12. Mercer and Mercer 1977, 7.
13. Johns 1973, 272.
14. Price 1991, 203.
15. Houston 1999, 203.
16. Drake 1998, 165.
17. Aventajado and Montelibano 2004.
18. Nance 2016, 124.
19. "Syria Conflict," September 10, 2014.
20. Levy, November 8, 2001.
21. Ibid.
22. Sayigh 2001.
23. Barghouti, January 15, 2002.
24. Machlis, June 16, 2002.
25. Marcus and Crook, July 14, 2004.
26. Tanner, April 4, 2002.
27. Stork 2002, 84.
28. START 2003.
29. Fletcher 2005.
30. Zelkovitz 2008.
31. Usher 2003, 28, 31, 43.
32. Pearlman 2011, 168.
33. International Crisis Group 2004, 26.
34. Stork 2002, 63, 82, 84.
35. Usher 2004.
36. Pearlman 2011, 177.
37. Bishara 2003, 48.
38. Stork 2002, 84.
39. Hammami and Hilal 2001, 41.
40. Barnea, September 2, 2001.
41. Marcus and Crook, July 14, 2004.
42. Usher 2003.
43. McGreal, December 22, 2002.
44. *Al-Hayat*, May 30, 2002.
45. Allen 2002.
46. Pearlman 2011, 169.
47. Bennet, July 7, 2004.
48. International Crisis Group 2004, i.
49. Greenberg, May 9. 2002.
50. Margalit 2003, 37.
51. Moghadam 2003, 73.
52. Sayigh 2001, 53.

53. Prudori, August 17, 2010.
54. "Taliban calls on fighters," November 5, 2011.
55. Rassler et al. 2012.
56. Kleponis 2010, 47.
57. Ibid. 13.
58. Dupee et al. 2009.
59. Clark and Osman, April 22, 2015.
60. Roggio, May 2, 2012.
61. Kleponis 2010.
62. UN News Centre, February 8, 2004.
63. Clark, October 19, 2011.
64. "The Taliban: A CFR InfoGuide."
65. Johnson and Dupee 2012.
66. Roggio, November 6, 2011.
67. United Nations Assistance Mission," 2014, 32.
68. Clark, October 19, 2011.
69. International Crisis Group 2004, 29, 33.
70. UN News Centre, February 8, 2004.
71. Magnowski, November 8, 2011.
72. Azam and Rosenberg, May 31, 2013.
73. Ibid.
74. Gopal 2013, 58.
75. Van Linschoten and Kuehn 2011.
76. Joseph 2017, 89.
77. Byman 2013.
78. Miglani, August 11, 2009.
79. Shahzad and Dozier, September 14, 2010.
80. Byman 2009, 19.
81. Schmitt, June 7, 2012.
82. Russell, June 14, 2011.
83. Michaels, January 24, 2012.
84. Abrahms and Potter 2015.
85. See ibid. for more discussion of the methods and results.
86. Johnston and Sarbahi 2016.
87. Zussman and Zussman 2006.
88. These results are adapted from Abrahms and Mierau, 2017.
89. Johnston and Sarbahi 2016.
90. See Abrahms and Mierau 2017 for a more detailed explanation of the empirical approach and results.

Rule #3
Branding to Win

12

Denial of Organizational Involvement

The first two parts of this book have shown that smart leaders recognize the costs of terrorism and take active measures to prevent it. These rules for rebels go a long way toward improving the tactical choices of the group and hence the chances of victory. But let's not overstate the power of the leader to impose task cohesion in the ranks. As we've seen, civilians are still struck about 15 percent of the time even when the leader is smart enough to oppose terrorism and central- ize the group to limit it. In the best-run groups, some militants will still occa- sionally commit terrorism due to ignorance, stupidity, carelessness, bad luck, or for personal reasons like revenge. Regardless of the exact cause, civilian attacks are a public relations fiasco for the organization because the perpetrators are typically seen as immoral extremists who need to be crushed rather than joined or appeased. Political psychologists affirm that images matter in international relations; how actors are perceived fundamentally affects their prospects of political success.[1] As Saul Alinsky eloquently put it, "All effective actions require the passport of morality."[2] Like the head of any organization, militant leaders must minimize the fallout to its reputation whenever members blunder. "When a face has been threatened, face-work must be done," the sociologist Erving Goffman reminds us.[3] So, what should a leader do when members threaten the reputation of the group with terrorism? How can he save face to salvage the cause? Militant leaders have more control over the image of their group than its actual behavior. The third rule for rebels is to brand the organization as moderate even when members act otherwise. In practice, this means denying organiza- tional involvement (DOI) in civilian attacks. This chapter explains the scientific basis of DOI and its pervasiveness in militant groups with smart leaders.

The Value of Denial

Militant leaders must be savvy with image restoration as it's known in com- munication studies. Image restoration is essential whenever a person or entity is accused or even suspected of having done something really bad.[4] The psychologist Barry Schlenker observes that "defenses of innocence" are critical whenever "events have undesirable implications for the identity-relevant images actors have claimed or desire to claim in front of real or imagined audiences."[5] For this reason, denials are widespread in all walks of life. Most investment bank CEOs responded to the 2008 stock market crash with categorical denials of responsibility.[6] Only about 30 percent of medical errors are disclosed to patients.[7] Reputational research finds that denial is a standard reaction for blame mitigation. "A common response to charges of misconduct is simply to deny any and all allegations," William Benoit notes in his seminal work on image restoration.[8] Other communication scholars like Lee Ware and Wil Linkugel also list denial as among the most common restoration strategies.[9] In their study of juvenile delinquency, for example, the sociologists Gresham Sykes and David Matza discovered that "denial of responsibility" is the most frequent approach to save face.[10]

Denials are pervasive in international relations, too. Japan denied the Imperial Army's gross mistreatment of Korean civilians during World War Two, including the sexual enslavement of countless "comfort women." Sakurai Shin, a member of the Murayama Cabinet, protested in August 1994: "I do not think Japan intended to wage a war of aggression ... It was thanks to Japan that most nations in Asia were able to throw off the shackles of colonial rule under European domination and to win independence." The Japanese government has commissioned numerous studies of this period proclaiming its innocence. A thirty-five-volume publication called "Historical Research on Japanese Overseas Activities" dismissed accusations of Japanese colonial exploitation as "absurd" and "lamentable" and "defamatory."[11]

International relations scholars generally see denials as ineffective, Jennifer Lind remarks: "As many analysts have argued, denials of past aggression and atrocities do fuel distrust and elevate fears among former adversaries. Japan's unapologetic remembrance—for example, frequent denials by influential leaders and omissions from Japan's history textbooks—continues to poison relations with South Korea."[12] And yet, denials probably wouldn't be so pervasive if they lacked utility. Winston Churchill once said truth may be "so precious that she should always be attended by a bodyguard of lies."[13] Ronald Reagan survived the Iran–Contra affair by denying that he had authorized the diversion of money from Iranian arms sales to the Contras. The "Teflon President" left office about as popular with the American public as before the illicit deal was exposed.[14] John Murtha avoided prosecution and then won re-election in Congress after denying any wrongdoing in the so-called ABSCAM scandal of the early 1980s.[15] Richard Nixon went on to become president after his famous Checkers speech denied the misuse of campaign funds.[16] Clarence Thomas secured confirmation of his nomination to the Supreme Court by denying Anita Hill's credible accusations of sexual harassment.[17] More recently, Sen. Ted Cruz weathered embarrassing news stories about watching porn by denying that he "liked" a porn tweet.[18] As in politics, business leaders are no strangers to the value of denial. A July 1993 Pepsi scandal fizzled shortly after the company denied allegations of syringes turning up in cans throughout the United States. Pepsi sales fell 2 percent during the crisis but fully recovered within a month.[19] In January 2001, rumors swirled that Taco Bell served an ersatz "mystery meat" product in its tacos, chalupas, and gorditas. The company released a statement that the allegations were "absolutely wrong," which restored its consumer base and averted a PR disaster.[20] Admittedly, both Pepsi and Taco Bell were innocent of those scurrilous charges. But there's considerable scientific evidence that denial helps with image restoration even when innocence cannot be established.

The communication scholar, Kevin McClearey, conducted an experiment in which subjects read a set of charges against the moral character of a fictitious person. Subjects were then randomly assigned to treatment groups in which the fictitious person actively denied the charges or issued other types of apologias. The study found, "Denial strategies were the most effective both in repairing damage to perceived credibility and in reversing perceptions of culpability."[21] In a related experiment, the psychologists, Catherine Riordan, Nancy Marlin, and Catherine Gidwani, determined that "denials decreased perceived wrongness of the act."[22] In another experiment, subjects weighed in on a hypothetical scenario of a plane that landed safely after catching fire. In one condition, the fictitious plane company issued a statement of "no-comment." In the other condition, the company engaged in "minimization" to reduce the perceived severity of the crisis by redefining it as more trivial than reported in the press. The study concluded: "Participants who were presented with the no-comment response reported significantly more trust in the organization and judged the organization as having less crisis responsibility."[23] In sum, there's substantial evidence in both observational and experimental settings that denials can aid image restoration in a variety of contexts.

The Logic of DOI for Militant Groups

Communication scholars emphasize that denial can help image restoration for a simple reason: "If the audience accepts the claim that an accusation is false, damage to the accused's reputation from that attack should be diminished, if not eradicated."[24] DOI may be either active or passive. In the former, the leaders assert the innocence of the organization. In the latter, they merely refrain from attributing to it guilt.[25] These insights have important implications for how militant leaders should respond when members defy their targeting guidelines with image-threatening attacks against civilians. Leaders should try to DOI whenever members commit these counterproductive attacks by disclaiming them or at least withholding organizational credit.

Recently, I interviewed a senior member of the Animal Liberation Front (ALF) to make sense of an apparent puzzle. Although decentralized, this group has seemingly exercised strong task cohesion in limiting the loss of innocent life. I was curious how a networked organization could buck the worldwide trend with its targeting restraint. It's very simple, she explained to me: Whenever people commit attacks that undermine the cause of the group, the ALF simply disavows them because the perpetrators cease to be part of the group. This way, ALF preserves its brand as an advocate for all "sentient" life when wayward members threaten it.[26] The Earth Liberation Front (ELF) employs the same branding strategy. Like ALF, ELF is regarded as a "leaderless" resistance group and emphasizes to members the imperative "to take all necessary precautions against harming life." As Joseph Brown notes, attacks that deviate from these guidelines are "not considered an ELF action . . . since a life was injured."[27] Historically, many militant leaders have conditioned taking responsibility for attacks on their anticipated costs to the group. The Popular Front for the Liberation of Palestine-General Command, Fatah, and Red Brigades would withhold credit for attacks until they got positive press coverage.[28]

Many militant leaders have been reluctant to claim credit for civilian attacks due to the expected political fallout. The official position of the African National Congress was that it had nothing to do with the May 1988 attacks on amusement arcades, fast-food outlets, sports stadiums, and shopping centers around Johannesburg and Pretoria.[29] Zawahiri publicly pretended that the damaging reports of AQI attacks on civilians which he privately opposed were just "lies concocted by the mainstream media" to discredit the group.[30] Sheikh Naim Qassem, Hezbollah's deputy chief, swore in November 2005 that his group "has never been involved in or responsible for any of these [alleged terrorist] incidents."[31] To help brand itself as a defensively oriented resistance group, Hezbollah has had fictitious organizations—like Islamic Jihad, the Revolutionary Justice Organization, the Oppressed on Earth, the Holy Fighters for Freedom, and the Defense of the Free People—place calls to media outlets taking responsibility.[32] In July 2014, militant Islamists shot dead a Muslim woman near the southern Somali town of Hosingow for refusing to wear a veil. To soften its image, an al-Shabaab spokesman denied the group had killed the woman. A *BBC* analyst noted, "Al-Shabaab wants to distance itself from the shooting because it is likely to provoke a strong public reaction."[33] In October 2017, a suicide bomber carried out the largest terrorist attack in Somali history when he detonated a truck packed with explosives in the streets of Mogadishu. As could be expected, thousands of Somalis took to the streets to demonstrate against the loss of over three hundred innocent lives. No official credit claim was issued to mitigate the reputational costs to the group. As the *New York Times* noted at the time, "The attack could backfire on the Shabaab—and that may be one reason the group has not claimed responsibility."[34] Leaders of the Al Qaeda affiliate in Syria likewise engaged in DOI for a March 2017 suicide bombing of a restaurant in Damascus, insisting that the group focused "only on military targets."[35] In August 2017, a Neo-Nazi named James Alex Fields drove his Dodge Challenger into a crowd of protesters in Charlottesville, Virginia. Hours before the lethal car-ramming, he had been photographed brandishing a shield emblazoned with a white supremacist emblem and other insignia of Vanguard America. As the pictures of Fields toting Vanguard America items circulated, the hate group distanced itself from the suspect over Twitter: "The driver of the vehicle that hit counter protesters today was, in no way, a member of Vanguard America. All our members had been safely evacuated by the time of the incident. The shields seen do not denote membership, nor does the white shirt. The shields were freely handed out to anyone in attendance."[36] Boko Haram leaders are also suspected of denying attacks "typically against civilian targets," according to UNICEF spokeswoman, Marixie Mercado.[37]

In Chapter 11, I detailed how the Taliban suffers from a severe principal–agent problem, as members often commit attacks on their own initiative at odds with targeting guidelines. Although unable to control which targets their operatives strike, the leadership can exert more influence over which attacks are claimed. Taliban leaders frequently practice DOI to brand the organization as more moderate than its targeting behavior. The leadership has been comparatively eager to take credit for attacks on military and other government targets. For instance, the Taliban "quickly claimed responsibility" when operatives ambushed Mohammad Qasim Fahim, leader of the alliance that toppled the Taliban in 2001, on a road in northern Kunduz in July 2009.[38] Not only does the leadership publicly celebrate such selective attacks, it even claims credit for those committed by other organizations, such as when the Haqqani network has struck Afghan and NATO installations in Khost, Paktia, or Paktika.[39] By contrast, the Taliban released the following statement when operatives defied their orders by attacking the International Committee of the Red Cross in Jalalabad: "The Islamic Emirate of Afghanistan wants to clarify to everyone that it was neither behind the May 29th attack on the I.C.R.C. office in Jalalabad city nor does it support such attacks."[40] The United Nations Assistance Mission in Afghanistan (UNAMA) affirms that Taliban attack denials are "frequently issued following civilian casualty incidents… perhaps highlighting the Taliban's continuous interest in gaining the Afghan people's support."[41] The governor of Farah Province also remarked: "Whenever there are civilian casualties, the Taliban deny responsibility."[42] After a Taliban attack on a Kandahar wedding for which the group denied responsibility, Radio Free Europe reported that the leaders "routinely deny causing civilian casualties."[43] When a bank was blown up in Jalalabad, France 24 anticipated that the Taliban would DOI as the leaders "rarely claim attacks that kill large groups of civilians."[44] The French news network pointed out that the Taliban withheld credit for bombing a wedding reception in the northern Afghan town of Aybak because "The group often distances itself from attacks with high civilian death tolls."[45] According to information minister, Mian Iftikhar Hussain, Taliban leaders eschew credit for terrorism because "They are desperate to wash their tainted image among the public."[46] In fact, the leadership is known to reverse its public stance upon discovering an attack harmed civilian targets rather than military ones.[47] Instead of taking credit for civilian attacks, the leaders try to attribute them to government forces. In February 2014, for instance, UNAMA published a detailed report on civilian casualties in Afghanistan. Of the 8,614 to occur in the previous year, 6,374 or 74 percent were assessed as Taliban perpetrated.[48] Predictably, though, the Taliban leadership refused ownership of these attacks and asserted that "civilian casualties are caused by the enemy itself" and that "the enemy is responsible for most incidents of civilian losses."[49] Its spokesman protested that such reports linking Taliban fighters to civilian casualties in Afghanistan are "propaganda," "far from reality," and "lies, all lies" intended to "cover up the blatant crimes of the Pentagon."[50] Some militant leaders, such as those in the Taliban, are apparently smart enough to DOI when civilians are killed to mitigate the audience costs.

With the Global Terrorism Database, I've tested with Justin Conrad whether leaders are more likely to deny organizational responsibility when

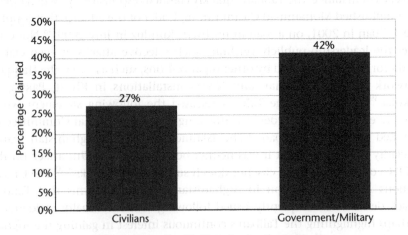

Figure 12.1. Credit Claiming by the Taliban

their members strike civilian targets versus military ones.[51] The GTD furnishes data on hundreds of militant groups throughout the world and includes variables on the presumed perpetrator of each attack, which group if any claimed credit, and the target of the violence. Figure 12.1 illustrates that the Taliban is indeed more likely to DOI when members strike politically risky targets. The Taliban claimed only 27 percent of its attacks against civilians compared to 42 percent against the military and other government targets from 2001 to 2011, a statistically significant difference.

Smart Leaders Practice DOI

This DOI pattern should hold for most—though not all—militant groups. In previous chapters, we've seen that leaders tend to be more terrorism-averse than the rank-and-file because they're more experienced, educated, intelligent, and dispassionate with weaker personal incentives for harming civilians due to their senior position in the organizational hierarchy. As such, we should find statistical evidence that militant groups throughout the world are generally more inclined to DOI when members strike civilian targets, especially as the leadership gains experience in asymmetric conflict and learns first-hand the costs of indiscriminate violence. We've also seen, however, that some leaders are strategic ignoramuses who never learn the costs of civilian attacks. Clearly, these leaders wouldn't be expected to DOI or even understand the value of doing so.

Table A.18 confirms that the leaders of militant groups around the world have disproportionately engaged in DOI when operatives struck civilian targets. Table A.19 shows that attacks on military targets are not only more likely to be claimed by the leadership, but to evoke competing claims of responsibility from multiple groups. The data thus support my contention that leaders tend to behave as rational actors who condition the decision to claim credit for an attack on its expected political return. For both sets of tests, the differences in means across the two targeting categories is statistically significant at the 0.01 level, meaning the odds of this finding occurring by chance is less than 1 percent. Although preliminary, these results suggest that militant leaders often have some inkling of the importance of image restoration, given the costs of civilian attacks to the reputation of the group.

Logistic regression analysis can help us to isolate the independent effects of target selection on DOI by controlling for alternative factors like the ideology of the group. Ideology is important to take into account because it supposedly influences whether a group claims credit for attacks. It's said that religious groups—especially Islamist ones—are less likely to claim organizational responsibility for violence because their only audience is god.[52] Table A.20 presents the regression results.[53] In all of the models, target selection is by far the most important factor in terms of whether militant groups around the world have claimed credit for their attacks. Holding the ten control variables constant at their mean values, civilian attacks are 41 percent less likely to be claimed by the leadership, demonstrating the substantial influence of target-ing type in its decision-making calculus.[54] The odds of this effect from target selection occurring purely by chance is less than 1 in a 1000. By contrast, religion has no statistically significant effect on the odds a group will claim or deny responsibility for its attacks. Although widely presumed to depress credit-claiming rates, Islamist groups are actually no less likely to assume responsibility for their attacks. Notably, groups with larger membership sizes are significantly less likely to claim credit for attacks. This makes sense based on my principal–agent argument because leaders lose control over the tactical choices of members as their numbers grow, requiring a higher rate of DOI for image restoration.

Together, these results provide additional evidence that militant leaders are generally more strategic than the rank-and-file. When members strike civil-ians, many leaders are wise enough to minimize the fallout by disavowing organizational responsibility for these image-threatening attacks. The results are robust to all sorts of specifications. Robust standard errors are clustered on the group to help account for correlations across attacks by the same group. Fixed country and time effects are applied across models. And the negative effects of civilian attacks on credit claiming rates hold with or without the inclusion of the Taliban in the sample. With over 1,500 attacks during this time-span, the Taliban is an outlier in terms of militancy. To ensure this group isn't driving the results, Table A.21 presents them when excluding the Taliban from the larger sample. Civilian attacks continue to significantly depress credit claiming, a testament to the fact many militant leaders have some appreci-ation of the value of moderate branding.

The fact that many militant leaders DOI civilian attacks isn't by itself evidence of a smart strategy. If wiser leaders take branding more seriously, we would expect them to DOI a larger portion of attacks as the costs become even clearer over time. To test my argument about the wisdom of DOI, I analyze the effect of time on two main dependent variables: the proportion of attacks for which the group claims credit and the proportion of civilian attacks for which the group claims credit. In Table A.22, I regress these dependent variables against the age of the group since its first recorded attack.[55] For each dependent variable, I present the results for all groups and only those that persisted for at least ten years to give the leaders more time to learn the value of DOI. Evidently, leaders become more careful about appear-ing too extreme with greater exposure to asymmetric conflict. As Models 1 and

2 reveal, groups are significantly less likely to claim credit over time and the results strengthen for those that survived at least a decade. Models 3 and 4 demonstrate that the percentage of civilian attacks for which groups claim credit also decreases over time, especially after a decade of fighting experience.[56] The wisdom of moderate branding through DOI is at least partly a learned behavior. Over time, leaders gain knowledge about how to advance the cause even when their members subvert it.

Of course, some leaders are incorrigible. They never learn. They're strategic dunces who refuse to follow any of the rules for rebels. An example from the Global Terrorism Datatbase is Al Qaeda in Iraq, notorious even in the upper echelon of its parent group for idiocy at the helm. Unlike other groups in the GTD, AQI was no more or less likely to claim credit when members attacked civilian or military targets. For these people any kind of violence is welcome. Clueless about the hazards of civilian attacks, Zarqawi and co. naturally didn't know any better to conceal them. With ISIS, the apple didn't fall far from the tree. Like its forebears, ISIS leaders eagerly claimed credit for almost all its terrorist attacks. This was the case even before AQI formally transitioned into ISIS. In August 2011, for example, Baghdadi gladly claimed organizational credit when suicide bombers in Mosul killed seventy people. And he also claimed credit for the December car bombings in a dozen Baghdad neighborhoods that killed at least sixty-eight people.[57]

Baghdadi learned little from the AQI experience. As leader of the Islamic State, he went to great lengths to brand the organization as even more extreme. July 2016 was a busy but otherwise unexceptional month for ISIS before it self-destructed. The string of attacks started in Nice on Bastille Day, when a Tunisian resident of France, named Mohamed Lahouaiej-Bouhlel, drove a 19-ton truck into a large crowd that was watching fireworks on the Promenade des Anglais, killing eighty-five people and injuring hundreds of others.[58] Islamic State proudly took responsibility for the carnage, issuing a statement thirty-six hours later praising the attacker as "a soldier of the Islamic State" who fulfilled his duty to indiscriminately kill bystanders.[59] Four days after the Bastille Day attack, Riaz Khan Ahmadzai took a machete to a family of four on a train near the German city of Wuerzburg. The next day, ISIS' online *Amaq* news agency released a statement claiming the seventeen-year-old asylum-seeker from Afghanistan was "one of the fighters of the Islamic State."[60] The following week, on July 24, a twenty-seven year old Syrian refugee, who had been denied asylum in Germany, blew himself up outside a wine bar in the city of Ansbach, wounding fifteen people. Before his backpack prematurely exploded, Mohammad Daleel had recorded a cell phone video pledging allegiance to Baghdadi, warning that Germans "won't be able to sleep peacefully anymore." *Amaq* swiftly confirmed that "Mohammad D." was indeed "one of the soldiers of the Islamic State."[61] A couple of days later, two soulless monsters ambushed a morning mass in a working-class village northwest of Paris, slitting the throat of the eighty-five-year-old parish priest, Father Jacques Hamel, while he was leading prayers. Of course, *Amaq* wasted no time to pronounce these "soldiers of the caliphate" as its own.[62]

ISIS leaders also claimed the October 31, 2015 downing of Metrojet Flight 7K9268 above the Sinai, a pair of suicide bombings two weeks later outside a mosque and bakery in a commercial district outside Beirut, and the Paris attacks the following day at the Stade de France in Saint-Denis, the Bataclan theatre, and some randomly chosen cafes and restaurants along the way. No attack has been beneath the group to claim. In fact, ISIS has occasionally claimed credit for mass casualty attacks on civilians when the perpetrator wasn't even inspired by the group, let alone directed by it. For instance, ISIS claimed credit for the 2017 attacks in Manila and Vegas by perpetrators who weren't even Muslims.[63] According to pundits in the media, "There's a hefty strategic logic behind every act of terrorism it claims."[64] This is true of many groups, but not ISIS. Like AQI, ISIS doesn't know to DOI. Their leaders have been too stupid to see that civilian attacks backfire, restrain members from committing them, and conceal organizational responsibility for the carnage, thereby ensuring the world coalesced against them. In Chapter 13, I'll explain what smart leaders do to brand the organization as moderate when DOI isn't a realistic strategic option for image restoration.

Notes

1. Holsti 1962; Jervis 1989.
2. Alinsky 1971, 44.
3. Goffman 1972, 27.
4. Benoit 1995, 67.
5. Schlenker 1980, 137.
6. Tahmincioglu, 20 October 2008.
7. "When Saying Sorry is the Right Thing to Communicate."
8. Benoit 1995, 3.
9. Ware and Linkugel 1973.
10. Sykes and Matza 1957.
11. Lind 2008, 25, 36.
12. Ibid. 3.
13. Nolan 2015, 35.
14. Lanoue 1989, 482.
15. Lovely, May 25, 2010.
16. Ryan 1988.
17. Benoit 1995, 161.
18. O'Keefe and Selk, September 13, 2017.

19. Bhasin, May 26, 2011.
20. Ibid.; Barclay, April 19, 2011.
21. McClearey 1983.
22. Riordan et al. 1988.
23. Lee 2004, 613.
24. Benoit 1995, 3.
25. Lee 2004.
26. E-mail interview with Ann Berlin of the Animal Liberation Front, May 13, 2017.
27. Brown 2017.
28. Cronin 2011, 108; Victor 2003, 19; Moss 1989, 58.
29. Battersby, August 21, 1988.
30. Lahoud 2010, 222.
31. Qassem 2010, 232.
32. Levitt 2013, 34.
33. "Somali woman killed," 30 July 2014.
34. Mohamed, Schmitt, and Ibrahim, October 15, 2017.
35. Amarasingam 2017.
36. Hensley, August 13, 2017.
37. *Associated Press*, August 22, 2017.
38. Afghan, July 26, 2009.
39. Ardolino and Roggio 2012.
40. Ahmed and Rosenberg, May 31, 2013.
41. United Nations Assistance Mission 2014, 32.
42. Roggio, November 20, 2009.
43. Blua, June 10, 2010.
44. "Suicide Attack in Afghanistan," April 8, 2015.
45. "Top MP Killed in Suicide Attack," July 14, 2012.
46. Yusufzai, July 23, 2011.
47. "Suicide blast slays 40," June 10, 2010.
48. United Nations Assistance Mission 2014.
49. Omar, August 6, 2013.
50. Prudori, August 17, 2010.
51. Abrahms and Conrad 2017.
52. Lesser et al. 1999. For an excellent study on credit claiming, see Hoffman 2010.
53. For a more detailed explanation, see Abrahms and Conrad 2017. The difference between Model 1 and Model 2 is the measure used to capture the competitiveness of the political environment. Model 1 employs the GTD variable for multiparty conflict; in Model 2, a count of organizations is used. Models 3 and 4 are identical to the first two models, but include control variables from BAAD.
54. These substantive effects are calculated using the results from Model 2. The difference in predicted probabilities is significant at the 0.05 level.
55. Fixed effects for groups are also included.
56. See Abrahms, Ward, and Kennedy 2018 for an alternative interpretation of these results.
57. Gerges 2016, 146–7.

58. "Death Toll from France Truck Attack," August 4, 2016.
59. Bazzi, July 27, 2016.
60. Rothwell, Graham, and Henderson, July 19, 2016.
61. "Official ISIS statement," July 25, 2016.
62. Olive, July 26, 2016.
63. Macaraign, June 4, 2017.
64. Dearden, October 6, 2017.

13

Denial of Principal Intent

The previous chapter showed that smart leaders take public relations seriously by denying organizational involvement in civilian attacks. But let's be real—DOI isn't always a realistic option. Sometimes, rebels are caught red-handed. What should leaders do when the culpability of the organization is undeniable? The key, then, isn't DOI but what I call DPI, or Denial of Principal Intent. With DPI, the leader acknowledges that his organization committed the terrorist attack, but denies that it reflects his intentions or the mission of the group more generally. In DPI, the leader admits something bad happened, but not by design. This PR move is critical because, as the experiment revealed in Chapter 4, people tend to infer the extremeness of an actor's intentions directly from the extremeness of his tactics in international relations. Victory will remain elusive if people around the world think the group is bent on slaughtering them. So, the next best approach after DOI is DPI to underscore that the extreme behavior of some agents represents neither the intentions of their principal nor the membership as a whole.

Luckily for rebels, there's a science to DPI as with DOI. This chapter prescribes an "account" for rebel leaders to restore the image of the group when its hand in terrorism is undeniable. Within social psychology, accounts are understood as self-serving public responses to mitigate the tendency of others to reach undesirable conclusions about the character of the offender.[1] Accounts are essential for offenders to limit damage to their reputations in face-threatening situations.[2] The field of communication identifies several ways for people to restore their image after an offense has been committed either by them or in their name. This chapter describes the most relevant accounts and how smart militant leaders apply them to distance their organization from the reputational fallout of terrorism. Although these accounts are cast in the academic literature as discrete approaches to image restoration, the shrewdest leaders combine them when organizational involvement in an offense is indisputable.

Image Restoration Account Strategies

Research in communication finds that *apologizing* is perhaps the most common account when blame is inescapable.[3] As William Benoit notes, "A defensive strategy for dealing with charges of wrongdoing that cannot be denied is to apologize for misconduct."[4] An apology is an expression of remorse or what communication scholars describe as a form of mortification for image restorative defense.[5] Notably, apologies need not take the form of a full concession. Concessions express regret, but go further than apologies by acknowledging the predicament as defined by others and assuming total responsibility for the offense.[6] Public apologies are commonplace with businesses facing PR disasters, such as when General Motors sold faulty ignition switches, the Apple Maps app didn't work, and Odwalla apple juice became E.coli-ridden. Surely, the CEOs of these companies would have preferred a DOI approach if organizational responsibility were deniable. But the CEOs proffered an apology instead to show their disproval of the negative outcome, highlight the goodwill of their companies, and thereby retain support. Public apologies are also commonplace in international relations, especially since the end of World War Two. Germans have practiced so much public atonement for their Nazi sins that they developed a big word for it—*Vergangenheitsbewältigung*.[7] In a recent book on this topic, the political scientist, Loramy Gerstbauer, remarks, "The idea of contrition in international relations is relatively new territory."[8] Jennifer Lind likewise observes, "Apologies for human rights violations have grown increasingly common."[9] The human rights scholar Elazar Barkan points out, "Admitting responsibility and guilt for historical injustices . . . has become a liberal marker of . . . strength rather than shame."[10] Ronald Reagan apologized on behalf of the United States for interning Japanese citizens during the war.[11] And Russian President, Boris Yeltsin, apologized for keeping Japanese prisoners of war as slave labor years after the war ended.[12] The offense need not be severe, such as when President George H. W. Bush apologized for vomiting in the lap of Japanese Prime Minister Kiichi Miyazawa at a 1992 banquet.[13] Apologies from the business world to international politics help leaders to project a positive image of themselves and the entities they represent.

Scapegoating is another standard account strategy to distance oneself from a perceived wrong.[14] Here, the person tries to shift the blame to others either entirely or through a diffusion of responsibility.[15] Scapegoating is a type of excuse. The philosopher, John Austin, describes excuses as an acknowledgement of an offense due to extenuating circumstances allegedly reducing personal culpability for the act.[16] The scapegoat excuse pleads the "defense of defeasibility," whereby the act is depicted as an unfortunate accident that happened for reasons beyond one's control.[17] The bible tells us that scapegoating is even older than apologies. Adam didn't apologize to god. He blamed Eve for eating the apple. During Yom Kippur, the sins of the Israelites were passed along to an unlucky goat who was thrown off a cliff. And in Greek mythology, the first woman was a scapegoat named Pandora. More recently, Presidential candidate, Hillary Clinton, went the time-tested scapegoating route when she blamed former Secretary of State, Colin Powell, for allegedly encouraging her to use a private e-mail server. Volkswagen CEO, Michael Horn, blamed the cars' engine software on a rogue engineer. Yahoo CEO, Scott Thompson, blamed inaccuracies in his CV on a recruitment firm.[18] And AT&T pinned the interruption in its long-distance service in New York City on "lower-level workers."[19] Like apologies, this image restoration tactic won't completely absolve the leader from blame but can signal positive intent after a fault had been committed.

Corrective action is another stock account strategy to signal goodwill. It represents the greatest acceptance of responsibility because the offender not only acknowledges the problem, but takes public measures to prevent future ones.[20] In business and politics, corrective action often means publicly punishing employees for their alleged blameworthiness, such as when Jamie Dimon sacked a rogue trader known as The Whale for JPMorgan trading losses and when President George H. W. Bush fired a campaign official after getting busted for digging up dirt on Bill Clinton.[21] Such face-saving measures can help the leader move on from a public relations fiasco by signaling that he didn't intend for the offense to happen and has taken steps to ensure it won't happen again. Whereas corrective action underscores the seriousness of an offense, *justification* attempts to redefine it as less severe to reduce the ill-feeling associated with the act.[22] Justifications try to minimize the perceived transgression by convincing the audience that it's more normative, understandable, and acceptable than how it appears.[23] An example of justifying is the kid who tells his mother that he crashed her car, but only banged up the bumper. A final account strategy is known as *bolstering* whereby the offender tries to refocus attention away from the misconduct to aspects of him viewed more favorably by the audience.[24] With this image restoration tactic, the offender highlights his less egregious, even positive behaviors to mitigate the ramifications of his transgression.[25] Some communication scholars brook no distinction between justifying and bolstering.[26] As with justifying, bolstering tries to lessen the negative impact of an offense, but by drowning it out with reminders of better behavior. Here, the kid tells mom that he crashed her car, but not to forget that he got straight A's on his report card.

These accounts may seem cheesy, but experiments conducted since the 1970s have consistently found that they work at image restoration. In one experiment, for example, subjects were presented with a vignette of a boy tripping a girl with a rope. The boy was silent in the control group, while apologizing in the treatment group. The psychologists found: "Expressions of remorse softened the perceived aggressiveness of the verbal boy, reduced the degree of intent attributed to him, lowered the probability associated with the predicted likelihood of repeating the action, and lightened the punishment assigned to him . . . The fact that remorse seems to work in lessening the negative reactions of others suggests that such behavior is usually rewarded."[27] In a similar experiment, the psychologists Kenichi Ohbuchi, Masuyo Kameda, and Nariyuki Agarie found that offenders were seen as less aggressive when they apologized for harming a girl in a variety of hypothetical scenarios.[28] The psychologists Marti Gonzales, Debra Manning, and Julie Haugen likewise demonstrated that student subjects imputed more benign intentions to the protagonist when he apologized for various face-threatening incidents.[29] In an experiment involving a hypothetical aviation mishap, a plane company was also seen as more safety-conscious when it apologized and took public actions to prevent future incidents.[30] In the most comprehensive experiment, William Benoit and Shirley Drew tested the effectiveness of multiple accounts on 202 students at Midwestern universities with five potentially face-threatening situations: failure to pick-up a roommate after work, mistake in a card game, refusal to clean a shared apartment, spilling a drink on a coat, and losing another's cassette tape. Subjects came away with a more favorable view of the alleged offender when informed that he apologized or scapegoated or pled the defeasibility defense or engaged in corrective action.[31] Experimental evidence suggests that bolstering also works. In another study, hundreds of students in Taiwan were presented with a hypothetical crisis situation of a gas explosion in a refinery plant that reportedly killed five bystanders when workers cut a pipeline. A treatment condition measured the effect on company perceptions when students were told that employees volunteered for community service. Inclusion of this positive information about the company significantly bolstered perceptions of its intentions toward the community.[32] In a quasi-experiment measuring the restorative effect of justifying the misconduct, Yi-Hui Huang presented participants with cases of Taiwanese political figures caught in extramarital affairs. Participants reported having a more favorable view of the politician when he tried to justify the affair by saying it happened with a longtime friend and "carelessness . . . resulted in this lapse."[33] The main upshot from this body of experimental research is that offenders aren't helpless to modulate perceptions of their blameworthiness. Accounts matter for shaping public understanding of culpability. They can be skillfully deployed to put a more sympathetic spin on the offender's role and disposition. Apologies can decouple him from his offense and convey more positive intentions. Excuses help offenders save face when the audience is told the incident was regrettable, accidental, understandable, out of character, and the fault of other people or factors beyond their own control, especially when corrective actions are taken to limit future transgressions.[34] The academic literature on accounts tends to describe them as separate face-saving approaches. But experimental studies suggest they work best in combination.[35]

Business executives typically employ a hybrid of account strategies in tandem to signal benevolent intent from the top and thereby repair their company image in a PR debacle. Texaco, for instance, invoked multiple accounts when six black employees sued the company in 1994 for discrimination. DOI wasn't an option because senior executives had been caught on tape deriding minority employees in racist terms. Texaco snapped into crisis management mode. Unable to deny organizational involvement for these allegations, the CEO Peter Biljur denied principal intent. He issued a public apology admitting personal embarrassment, blamed the apparent racist culture on the scapegoats caught on tape, took corrective action by suspending them, and bolstered the company image with a PR blitz underscoring its commitment to diversity and charitable works.[36] In 1999, Coca-Cola went into crisis communications mode when customers in Belgium and Northern France complained of nausea and dizziness after drinking the product. The executives released a statement apologizing in Belgian newspapers for the sicknesses, blaming them on flawed carbon dioxide used by one of their bottlers, and bolstered the company image by paying the medical bills and distributing coupons to four million Belgian homes.[37] In 2003, Cadbury stared down its own PR crisis when several of its chocolate bars in Mumbai got infested with worms. Cadbury responded by issuing a public apology, blaming the infestation on poor storage at retail stores, justifying the problem by highlighting that only a miniscule portion of its chocolate bars were tainted, and adopting worm-proof packaging to prevent repeat occurrences.[38] Burger King executives dealt with their own PR nightmare in 2008, when an employee known on MySpace as Mr. Unstable posted a video of himself bathing in the kitchen sink at the Xenia, Ohio franchise. Within days, the video had been viewed nearly a million times, attracting damaging international press coverage. Burger King execs responded by apologizing for the disgusting incident, blaming the rogue employee, publicly firing him, sterilizing the sink, and retraining the other staff in sanitation procedures.[39] In 2009, Domino's adopted similar crisis communications when a couple of employees in its Conover, North Carolina, franchise posted a prank video on YouTube blowing snot and farting in the food. President Patrick Doyle employed multiple restorative accounts to distance both him and his company from the horror. On YouTube, he released his own video, which apologized for the fiasco, denigrated the employees in question, divulged they had been fired, and engaged in both justifying and bolstering by noting that these rogue individuals make up a tiny portion of Domino's workforce that propels the economy. As Doyle mentioned in his YouTube apology, "It sickens me to think that two individuals can impact our great system, where 125,000 men and women work for local business owners." Domino's spokesman Tim McIntyre released an additional statement to help restore the company image, "We got blindsided by two idiots with a video camera and an awful idea."[40] These account strategies help executives to restore the image of the company by denying principal intent and thereby signaling goodwill when organizational culpability is undeniable.

Terrorist Accounts

Smart militant leaders apply the same DPI accounts when forced to admit members harmed civilians. Militant leaders issue apologies more often than you might think. One study found in a sample of militant groups with GTD data that leaders have apologized for about 22 percent of attacks. Guess what the key variable is affecting the likelihood of an apology? In accordance with my argument throughout this book, target selection is the most important factor. Civilian casualties reportedly increase the odds four-fold of an apology on behalf of the group.[41] The IRA leadership, for example, initially denied organizational involvement in a December 1983 car-bombing outside the Harrods department store in central London that killed two civilians and seriously injured fourteen other bystanders. When DOI became an impossibility, the leaders took the DPI route by issuing an apology.[42] The IRA also apologized in 1987 when a bomb killed eleven bystanders at a memorial service for veterans in Northern Ireland. The public statement read: "The IRA admits responsibility for planting the bomb in Enniskillen yesterday which exploded with such catastrophic consequences. We deeply regret what occurred."[43] The weekly *Republican News* reiterated that the attack was a "monumental error."[44] Similarly, the Basque separatist group ETA couldn't escape responsibility for a 1987 car bomb at the Hipercor shopping center in Barcelona that killed twenty-one bystanders. The leadership issued a public apology when an estimated 750,000 people marched through Barcelona with banners declaring, "Catalonia Rejects Terrorism."[45] In 1999, Nicolas Rodriguez Bautista's Colombia National Liberation Army apologized after members kidnapped and killed worshippers at a church: "We beg forgiveness from the Church and all the faithful for this act."[46] In 2004, the leader of the Fatah-affiliated Popular Resistance Committee issued an apology when members shot at a vehicle in Gaza, killing an entire family including four young children. Jamal Abu-Samhadana stated in a post-attack interview: "When I heard that children were killed I was ashamed."[47] Similarly, Samir Masharawi and Sami Abu-Samhadana apologized in 2005 for civilian attacks inside Israeli cities, which the leaders of Fatah's Tanzim faction in Gaza characterized as "war-crimes."[48] In 2005, Prachanda apologized when

Nepalese rebels made a "grave mistake" in bombing a bus: "The incident was against our party policy and we offer serious self-criticism to the general public for the huge civilian casualties."[49] During its war against Israel in the summer of 2006, Hezbollah rocket-fire killed two Israeli-Arab children. To minimize the PR fallout, Nasrallah stated on *Al Jazeera*: "To the family that was hit in Nazareth . . . I apologize to this family . . . those who were killed in Nazareth, we consider them martyrs for Palestine."[50] Over the years, Al Qaeda leaders have increasingly issued apologies when members harmed civilians. Bin Laden famously apologized for Zarqawi's *takfirism*.[51] In 2009, the American Al Qaeda spokesman, Adam Gadahn, released a video message in English to "express our condolences to the families of the Muslim men, women, and children killed" in terrorist attacks in Afghanistan and Pakistan.[52] In 2013, Al Qaeda in the Arabian Peninsula struck a hospital that left dozens of people dead. Emir Qassim Al-Raimi released an apology over the Al-Malahim media outlet: "We confess to this mistake and fault. We offer our apologies and condolences to the families of the victims. We did not want your lost ones; we did not target them on purpose . . . It wronged us and pained us because we do not fight in this manner."[53] In 2014, a suicide bomber from the Lebanon-based Abdullah Azzam Brigades attacked a Shi'ite cultural center in Beirut and killed eight bystanders on the street. The Al Qaeda-linked group apologized over Twitter: "The Abdullah Azzam Brigades' operations do not target Shi'ites in general, or any other sect, and we always stress to our martyrdom-seekers to be cautious and abort an operation if they think it could kill others than those targeted."[54] In 2015, Nusra members entered a village in Idlib province and shot dead at least twenty Druze villagers. This indiscriminate attack came at a sensitive time in the Syrian conflict when the Al Qaeda affiliate was trying to brand itself as a more moderate Salafist alternative to ISIS. Via Twitter, the leadership extended an apology to the Druze community, saying it had "received with deep sorrow news of the incident."[55] In 2016, members of the U.S.-vetted Nour al-Din Zenki Movement in Syria beheaded a Palestinian kid. In a statement posted on its Facebook page, the rebel group stated that such abuses are "individual errors that represent neither our typical practices nor our general policies."[56] When civilians are struck, apologies like these help militant leaders to project more positive intentions both personally and on behalf of their group.

Militant leaders typically combine an apology with scapegoating to distance themselves from the offense. Essentially, the principal blames the unwanted attack on his agents. In their apology for the 1983 Harrods attack, for example, the IRA Army Council claimed it had "not authorized the attack," pinning blame on the operatives.[57] Similarly, the IRA said that the Enniskillen attack was "not sanctioned by the leadership."[58] Oliver Tambo, the Chairman of the African National Congress in exile, likewise attributed a 1984 bombing in Durban that killed four white civilians to an "inexcusably careless" fighter."[59] When apologizing on behalf of the National Liberation Army for the 1999 church attack, the leader blamed his fighters for their unfortunate target selection.[60] When an Al-Aqsa Martyrs Brigades operative shot three civilians at a Tel Aviv restaurant in 2002, the unit leader Mahmud Al-Titi said he "saw soldiers or guards next to the restaurant [which is why he shot in that direction] or maybe he didn't find soldiers or police and so attacked the closest target."[61] Regardless of the precise chain-of-events, the attack wasn't approved from the top. The leader of the Popular Resistance Committee likewise said that the assailants who shot at the family-packed car in Gaza mistook it for a military vehicle because "this road is used by IDF Jeeps."[62] In his apology for the bus attack, Prachanda said that operatives mistook it for a military vehicle.[63] The leader of Lashkar-e-Taiba, Hafiz Muhammad Saeed, likewise blamed the 2008 Mumbai attacks and other instances of violence against the population on "rogue elements within the group."[64] Using the same language, Taliban leaders blame face-threatening civilian attacks on unrepresentative "rogue elements" in the ranks.[65] In his apology, Al-Raimi likewise scapegoated the terrorist who attacked the Yemeni hospital: "We saw what the Yemeni channel broadcast: a gunman entering a hospital... We did not order him to do so, and we are not pleased with what he did."[66] Leaders in the Abdullah Azzam Brigades said the suicide bomber who attacked the Shi'ite cultural center in Beirut violated the operational guidelines "to be cautious and abort an operation if they think it could kill others than those targeted."[67] In their apology, Nusra leaders said the terrorists who struck the Druze village hadn't consulted their commanders and committed "clear violations" of targeting policy.[68] In their apology for beheading the Palestinian kid, the Zenki leadership stressed that "the boy's assailants had acted independently" of the Syrian rebel group.[69]

As the communication literature explains, scapegoating is a classic "defense of defeasibility" excuse in which leaders depict the face-threatening offense as an unfortunate accident that occurred for reasons outside their control. This is a common DPI strategy by militant leaders—to emphasize that the accidental killing of civilians happened under extenuating circumstances that somehow mitigate blameworthiness. For instance, the Enniskillen attack allegedly happened for unforeseeable reasons: "One of our units placed a remote control bomb... aimed at catching Crown forces personnel on patrol in connection with the Remembrance Day service but not during it. The bomb blew up without being triggered by our radio signal."[70] In the Abdullah Azzam Brigades' account, the attack on the cultural center regrettably resulted in civilian casualties because the shrapnel from the explosion travelled further than normal.[71] When members of the Popular Resistance Committee shot up the vehicle in Gaza, "They couldn't have known that there were children [in the car]."[72] The defeasibility defense sometimes places blame on the government instead of the attackers, as when leaders in the African National Congress, ETA, Irgun, and IRA blamed officials for failing to heed their warnings to evacuate civilians from the bombing area.[73] The implication of these accounts is that the terrorist attack wasn't the leader's fault or at least his intent.

Like CEOs, militant leaders often engage in corrective action to convey their opposition to an offense and commitment to preventing future ones. In Chapter 9, we saw that leaders frequently discipline members for tactical mistakes. Such punishments are doled out at least as much for image restoration as for restraining members. In the Zenki apology, for instance, the leader emphasized that the head-choppers would be brought before a disciplinary tribunal for their sickening act.[74] The IRA apology for the Enniskillen attack made a point of mentioning that the perpetrators would be subject to an "internal inquiry."[75] Prachanda stressed that the Nepalese rebels who blew up the passenger bus were suspended from the group.[76] When twenty Druze villagers were killed by members of Al Qaeda's Syria affiliate, the leadership not only scapegoated them, but made them stand trial before an Islamic court "to account for blood proven to have been spilt."[77] Such public punishments are a form of crisis communications to help militant leaders distance themselves and their other members from the PR fallout.

At the same time, militant leaders try to justify attacks in order to lessen their reputational costs to the group. The protest leader, Saul Alinsky, emphasized, "You do what you can with what you have and clothe it with moral garments."[78] To clothe attacks as more moral, militant leaders often contest their civilian nature. As the political scientists, Or Honig and Ariel Reichard, point out, "Terrorists try to excuse the killing of civilians . . . by claiming that the specific victims killed were not innocent civilians as their critics claim but rather were engaged in or contributed to the government's military efforts against it and hence constituted legitimate military targets."[79] In 1985, the Red Army Faction killed Edward Pimentel, an off-duty American soldier stationed in Germany. The murder drew a rebuke because he was, at that moment, a noncombatant. The RAF issued a communiqué emphasizing his status as a belligerent to justify the murder: "We shot Edward Pimentel, the air defense specialist, who was in the army of his own free will, and in the FRG [Federal Republic of Germany] for the last three months . . . the air base . . . stands at the center of the confrontation between the international struggle for liberation and imperialism—which is war—and, as such, so do all the soldiers who are there."[80] In 1992, an IRA roadside bomb in Omagh County destroyed a van carrying fourteen protestant workers. The Catholic community criticized the IRA for the indiscriminate violence. To temper the outcry, the IRA claimed the workers were legitimate targets because they were "collaborating" with the "forces of occupation" by re-building a British

Army base.[81] Palestinian terrorist leaders often justify attacks on Israeli noncombatants by describing them as settlers.[82] Alternatively, Israeli civilians are depicted as legitimate targets because "They are part of the total population, which is part of the army...From 18 on, they are soldiers, even if they have civilian clothes."[83] Similarly, a notorious Chechen militant Islamist liked to say all Russians are legitimate military targets: "They pay taxes. They give approval in word and in deed. They are all responsible."[84] When they assume organizational responsibility for attacks against civilians, Taliban leaders routinely say they weren't really civilian. For instance, a Taliban spokesman insisted that the twenty-seven laborers shot dead in October 2008 were secretly "Afghan National Army soldiers...traveling to Helmand wearing ordinary clothes."[85] After killing a number of civilians in June 2012, Taliban leaders insisted the martyrdom operation took out "foreign terrorists [NATO forces] and their cowardly local puppets [Afghan security forces]."[86] And when another suicide attacker killed nine Afghan civilians in Kajaki earlier that year, Taliban leaders declared that the operation instead killed seventeen troops.[87] Jihadist leaders often invoke the concept of *Qisas* to justify civilian attacks.[88] This law of equal retaliation from the Islamic penal code holds: "If the unbelievers have targeted Muslim women, children, and the elderly, it is permissible for Muslims to respond in kind and kill those similar to those whom the unbelievers killed."[89] In these ways, militant leaders try to justify the attack by denying not organizational responsibility, but the interpretation of it as terrorism.

In addition to justifying attacks, militant leaders engage in the closely related strategy of bolstering. Whereas justifying tries to make the alleged offense seem more acceptable, bolstering supplies additional positive information to distract from it and signal altruistic intentions. To this end, militant leaders call attention to social welfare projects like their provision of free schooling, medical care, and religious education. Even more importantly, militant leaders claim their intent is to weaken the enemy even if civilians are inadvertently harmed in this noble process. Public apologies therefore tend to pair regret for civilian attacks with appeals that the intended target was military. In his e-mailed apology to journalists for the botched bus attack, Prachanda stressed that "the mistake" happened in the course of "our fighters targeting the soldiers of the royal government."[90] After scapegoating operatives for attacking a family in Gaza, the leader of the Popular Resistance Committee noted that his men erred while going after legitimate "military targets."[91] The leader of Colombia's National Liberation Army said the church attack was unauthorized and regrettable, but intended to counter a "right-wing death squad" that had recently killed twenty-five peasants in their stronghold.[92] IRA apologies for civilian tragedies reiterated the goal of combating "British army troops" and the "Royal Ulster Constabulary."[93] In his apology for beheading a Palestinian boy, the Zenki leader reminded audiences that the mission is to combat the enemy "regime" of Bashar Assad.[94] The apology for inflicting collateral damage at the cultural center in Beirut emphasized that the Abdullah Azzam Brigade's "fight was against Iran and its ally Hezbollah, not Shi'ites in general."[95] Similarly, Nusra's apology for killing Nasaraa [Christian] bystanders in a March 2012 attack in Damascus emphasized that it was directed against the "air force security building" responsible for barrel bombing Syrians.[96] In these ways, militant leaders try to bolster the image of the group by underscoring that its intentions are inoffensive, even laudable.

Propaganda of the Deed

Propaganda videos are the primary vessel for militant leaders to bolster the image of the group. These days, videos often shape the image of the group even more than its actual behavior. Like a CEO, smart militant leaders carefully manage their brand by deploying propaganda videos as advertisements to maximize organizational appeal. In practice, this means bolstering the image of the group by projecting a moderate brand even when members act otherwise by attacking civilians.

Leaders of the Free Syrian Army, for example, have been careful to put a moderate face on their product to win public approval. Public relations are enshrined in the Code of Conduct. Article VI requires members to come across as moderate by subscribing to the following terms: "I will not participate in any public execution." Executions are fine, just not public ones because they risk scaring off supporters.[97] The so-called White Helmets likewise gained considerable local and external support by selling themselves as a moderate force for good in Syria. When members were caught on video participating in a public execution and subsequent cover-up, the leaders promptly suspended them to repackage the White Helmets as altruistic do-gooders for the Syrian people.[98] My colleagues and I at Northeastern University conducted a content analysis of propaganda videos to systematically assess the branding strategies of militant groups relative to their actual behavior.[99] Specifically, we compared the targeting choices of militant groups from the Global Terrorism Database to the attacks featured in their propaganda videos from the Intel-Center Database. Not only are propaganda videos significantly more likely to feature attacks against military targets than civilian ones, but the image presented by most groups systematically differs from their real-world targeting record. As you can see in Figure 13.1, even radical groups like Al Qaeda in the Islamic Maghreb, Al Shabaab, and the Taliban try to brand themselves as more

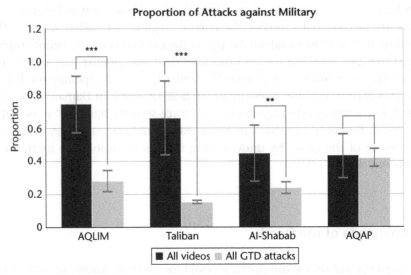

Figure 13.1. Image Versus Reality: Proportion of Attacks Against Military

moderate by featuring in their propaganda videos a larger proportion of attacks against military targets than they've historically committed.

Not all groups practice moderate branding. Al Qaeda in the Arabian Peninsula (AQAP), for example, is still very extreme in the violent images it presents to the world. In general, militant leaders seem to learn over time the value of moderate branding. When it comes to civilian attacks, DPI like DOI isn't static behavior. Just as leaders increasingly withhold organization credit for civilian attacks when culpability is deniable, leaders learn to downplay them in propaganda to bolster the image of their intentions. In May 2004, Zarqawi decapitated an American freelance radio-tower repairman named Nick Berg and then broadcast the atrocity in an online video. Bin Laden and Zawahiri saw the global backlash. By year's end, they had officially banned beheading in Al Qaeda propaganda videos.[100] In his October 2005 letter to Zarqawi, Zawahiri excoriated him for failing to learn the value of more moderate branding. "Among the things which the feelings of the Muslim populace who love and support you will never find palatable," Zawahiri said, "are the scenes of slaughtering hostages." If you need to kill a captive do it quietly "by bullet."[101] Zawahiri tried to impress upon Zarqawi that moderate branding is essential for the success of Al Qaeda in Iraq because "we are in a media battle in a race for the hearts and minds of our community."[102] But Zarqawi was too stupid to listen.

The leadership of the Al Qaeda affiliate in Syria got the message, however. As I explain in a recent exchange with Barbara Walter in *International Security*, Nusra attracted substantial local and international support by selling itself as a relatively moderate Salafist alternative to ISIS.[103] Skillful crafting of propaganda lay at the core of this branding strategy. Like the PR methods of Hezbollah and the Muslim Brotherhood, Nusra established its own media network called al-Manara al-Bayda (the White Minaret) to downplay its civilian attacks while publicizing its charitable deeds. In December 2012, for example, the Al Qaeda affiliate launched a video of members handing out bread to residents of Deir ez-Zor. What mattered wasn't so much that Nusra established its own relief department called Qism al-Ighatha to provide food and aid to the poorest, but that the leadership broadcast the hearts-and-minds initiative for potential supporters to see.[104] An August 2013 video bolstered the Nusra brand with footage of an Aleppo-based family fair during Eid celebrations, replete with members doling out Teletubbies and Spiderman dolls to grateful children.[105] Another online video depicts a newly acquired firetruck assisting with garbage collection in the city of Daraa, demonstrating Nusra's ability to run the country.[106] As the Syria war raged, Nusra shed its formal Al Qaeda affiliation altogether, changing its name to Jabhat Fateh al-Sham and then Hay'at Tahrir al-Sham to reap the benefits of more moderate branding. To be clear, none of these iterations of the group has ever been truly moderate. But the leaders have been smart enough to grasp the value of appearing so compared to both its own behavior and its obtuse Islamic State rival.

Like Hezbollah, Shi'ite militia leaders in Iraq have also become more astute about the perils of seeming too extreme. Abou Azrael a.k.a. "Iraq's Rambo" is a muscle-bound fighter for the Brigades of the Imam Ali notorious for meting out gory punishments and broadcasting them to the world. In 2014, Rambo was filmed decapitating some captives before burning their corpses. A 2015 snuff film shows him cutting up somebody's body with a sword. In 2017, "Rambo" was filmed dragging corpses around by the hair. This time, though, his leaders understood that broadcasting such atrocities risked eroding organizational support. The leadership cracked down on him and released this statement in June: "These acts were committed by an individual in his own name" and "the people who committed these acts will be prosecuted." Wassim Nasr, a journalist at France 24, noted that the about-face was a learned behavior to improve the brand of the group, as "Shiite militias don't want to present a violent image of themselves on the world stage."[107] Like the CEOs of Texaco, Coca-Cola, Cadbury, Burger King and Domino's, the militia leaders invoked a hybrid of restorative accounts to deny principal intent by apologizing for the offense, scapegoating it, and taking corrective action to protect the brand.

By contrast, the Islamic State brand is synonymous with indiscriminate carnage. The leadership doesn't just claim credit for wanton attacks against civilians. It brags about them. In doing so, the leaders have signaled not only

ISIS' involvement in terrorism but that the extreme behavior accurately reflects their intentions and the mission of the entire group. This disastrous PR strategy began even before the formal declaration of the Islamic State. In October 2013, a document in Aleppo extolled the practice of crucifying civilians.[108] This directive was included in a March 2014 compilation of *hudud* released in Aleppo.[109] By springtime, ISIS supporters were Tweeting images of crucifixions in towns across Syria.[110] This was just the start, though, of publicly celebrating unrestrained savagery. Over the next couple of years, ISIS leaders would broadcast tens of thousands of official propaganda products, including daily videos showcasing its members decapitating hapless victims with a knife, blowing their heads off with a detonation cord, burying civilians alive, chucking them from rooftops, setting them ablaze, drowning them in cages, and crushing them with tanks.[111] In some of these videos, the victims are children.[112] In other videos, they're parents appealing to the cameramen to spare them because they're civilians.[113] For ISIS, not all attacks are of equal propaganda value. The leaders prefer gruesome attacks against civilians, especially mass-casualty ones.[114] A July 2016 *Amaq* infographic lists "the most important attacks by fighters" in terms of their lethality around the world. Based on this rubric, the Paris killing spree that killed 200 was the best. Then came the Nice truck-ramming that killed eighty-four. The Orlando mass shooting at a nightclub was listed next because it took fifty-four lives. Almost as high on the list were the forty killed in Brussels, the thirty-five in Sousse, and the twenty-four stabbed to death in Dhaka.[115] By contrast, non-lethal attacks like the January 2016 shooting of a policeman in Philadelphia and the knifing of a Paris policeman went unacknowledged in *Dabiq*, *Rumiya*, and other official ISIS propaganda outlets.[116] To be a hero in ISIS requires mass killing, especially of innocent people.

To be fair, ISIS has engaged in softer forms of branding as well, with videos of its schools, hospitals, even a grand hotel in Mosul.[117] But such imagery is easy for non-specialists to miss precisely because the leadership has devoted so much effort to branding the group as the enemy of the human race. Not only does ISIS claim organizational responsibility for counterproductive attacks, but the leaders boast about them rather than apologizing, scapegoating, correcting, justifying, or bolstering the face-threatening behavior to minimize the fallout, as every competent CEO knows to do. Fed by an eager stable of think tank pundits, international media outlets from the *New York Times* to *CNN* repeatedly asserted without any scientific basis that this losing branding strategy was actually a winning one.[118] As an article in *War on the Rocks* concluded, "When it comes to strategic storytelling, the Islamic State truly has been unmatched—not only in terms of the quality of its output, but in quantity, too."[119] Reflecting the conventional wisdom, the authors claim that ISIS is the best ever at "image management."[120] Based on this profound misreading of human psychology and history, social media companies like Twitter and Facebook and YouTube got hammered for allowing their platforms to disseminate ISIS atrocities.[121] In reality, these companies should have been thanked—not condemned. After all, they helped ISIS to fall on its own sword by exposing its evil face. ISIS propaganda will continue to inspire some attacks around the world, the very kind that have ensured the demise of supporters and the caliphate itself.

Notes

1. Gonzales et al. 1992, 958.
2. Snyder 1985; Snyder and Higgins 1988.
3. Scott and Lyman 1968, 46; Lee 2004, 603.
4. Benoit 1995, 4.
5. Brinson and Benoit 1999.
6. Gonzales et al. 1992, 958.
7. Lind 2010, 100.
8. Gerstbauer 2017, v.
9. Lind 2010, 126.
10. Barkan 2011, xix.
11. Lewis, May 27, 2016.
12. "Japanese Get Apology," October 12, 1993.
13. Wines, January 9, 1992.
14. Lee 2004.
15. Schlenker 1980; Scott and Lyman 1968.
16. Austin 1964, 124.
17. Scott and Lyman 1968.
18. Schutte, March 9, 2016.
19. Benoit 1995, 27.
20. Lee 2004, 603.
21. Zamasky, March 15, 2013.
22. Benoit and Drew 1997, 156.
23. Gonzales et al. 1992, 958.
24. Ware and Linkugel 1973, 277.
25. Benoit and Drew 1997, 156.
26. Huang 2006.
27. Schwartz et al. 1978.
28. Ohbuchi et al. 1989, 219.
29. Gonzales et al. 1992, 967.
30. Lee 2004, 613.
31. Benoit and Drew 1997.
32. Wan 2004.
33. Huang 2006.

34. Riordan et al. 1988; Lee 2004; Gonzalez et al. 1992.
35. Huang 2006.
36. Bhasin, 26 May 2011; Fullerton et al. 2013.
37. Davies et al. 2002, 109.
38. Bhasin, May 26, 2011; "Worms Creep Out of Cadbury," 12 October 2003.
39. Marco, August 12, 2008.
40. Agnes, March 22, 2012.
41. Matesan and Berger 2017.
42. "IRA Says Harrods," December 30, 1993.
43. Matesan and Berger 2017, 388.
44. Raines, November 15, 1987.
45. Delaney, June 23, 1987.
46. "Rebel Leader Apologizes," June 7, 1999.
47. Eldar 2005, 186.
48. Honig and Reichard 2017, 10.
49. *Associated Press*, June 8, 2005.
50. Honig and Reichard 2017, 9.
51. Scheuer 2011, 148.
52. "Al Qaeda offers condolences," December 13, 2009.
53. *CNN*, December 22, 2013.
54. *Reuters*, March 8, 2014.
55. Soguel, 15, June 2015.
56. Armstrong and Alkshali, July 20, 2016.
57. O'Day 1997, 20.
58. Matesan and Berger 2017, 388.
59. Sparks 1991, 247.
60. "Rebel Leader Apologizes," 7 June 1999.
61. Daraghmeh, March 6, 2002.
62. Eldar 2005, 186.
63. "Nepalese rebels apologize," June 7, 2005.
64. Subrahmanian et al. 2013, 34.
65. *New York Times*, May 9, 2011.
66. *CNN*, December 22, 2013.
67. *Reuters*, March 8, 2014.
68. Soguel, June 15, 2015.
69. Byrd 2016.
70. Matesan and Berger 2017, 388.
71. *Reuters*, March 8, 2014.
72. Eldar 2005, 186.
73. See, for example, Collins and McGovern 1998, 120.
74. Byrd 2016.
75. Matesan and Berger 2017, 388.
76. "Nepalese rebels apologize," June 7, 2005.
77. "Syria's al-Qaeda affiliate says it regrets,' June 13, 2015.
78. Alinsky 1971, 36.

79. Honig and Reichard 2017, 16.
80. Ibid.
81. Ibid.
82. Ibid.
83. Safieddine, October 23, 2004.
84. Richardson 2007, 6.
85. Naadem, October 19, 2008.
86. Roggio, June 6, 2012.
87. Ibid., January 19, 2012.
88. Bassiouni 2014, 140.
89. "A statement from al-Qaeda," April 24, 2004.
90. "Nepalese rebels apologize," June 7, 2005.
91. Eldar 2005, 186.
92. "Rebel leader apologizes," June 7, 1999.
93. Raines, November 15, 1987.
94. Armstrong and Alkshali, July 20, 2016.
95. *Reuters*, March 8, 2014.
96. Gerges 2016, 184.
97. "Syria: Code of conduct."
98. "Syria's white helmets," May 19, 2017.
99. Abrahms, Beauchamp, and Mroszczyk 2017.
100. Nance 2016, 393.
101. Scheuer 2011, 144.
102. Al-Zawahiri 2005.
103. Abrahms 2018.
104. Gerges 2017, 187.
105. Fisher, August 13, 2013.
106. Gerges 2016, 187.
107. "Iraq's Rambo," June 15, 2017.
108. "Statement from the governor of Aleppo," 2013.
109. "List of Hudud Punishments," 2014.
110. Siegel, April 30, 2014.
111. Nance 2016, 256; Winter, July 2015.
112. Weiss and Hassan 2015, 171.
113. Ibid.
114. Brown 2017.
115. "Official ISIS Statement," 25 July 2016.
116. Brown 2017.
117. Gerges 2016, 266.
118. For a small sample, see Lister 2016, 4; Barnard and MacFarquhar 2015; Engel 2015; Goldsmith 2015; Weiss and Hassan 2015, xv, 171; Koerner 2016.
119. Clarke and Winter, August 17, 2017.
120. Ibid.
121. Nance 403.

Conclusion: The Future of Terrorism

Research on militant groups neglects the role of leaders. This omission is striking given the emphasis governments place on capturing and killing them. My book is the first to show how militant leaders affect the caliber of decisions and hence the prospects of victory. Of course, smart leaders aren't always successful. But successful leaders must be smart. They know which targets are costly, how to avoid them, and what to do when members don't. Each rule is based on the value of civilian restraint. Contrary to the conventional wisdom in think tank and academic circles, the savvy leader understands that terrorizing innocent people is counterproductive, so he limits the occurrence, and then distances the organization whenever members fail to comply. Like other militant groups with stupid leaders, ISIS paid a steep, albeit foreseeable, price for inverting these rules for rebels. Its playbook of broadcasting indiscriminate attacks instilled terror around the world, but quickly united it against the group. Militant leaders have a choice. They can maximize terror as political losers or forgo the terror and possibly win. The good news is that moderation pays. The bad news is that extremism will nonetheless continue.

What to Expect

Terrorism will continue so long as leaders believe that it works. This misperception about terrorism endures notwithstanding its abysmal political track record. Such disconnects between perception and reality are an inescapable aspect of the human condition. Cognitive psychologists have established that humans are prone to the "availability heuristic" in which the salient is mistaken for the statistical.[1] That's a real issue when evaluating the effectiveness of terrorism because "terrorism's successes have been more visible than its failures."[2] The sociologist, Randall Collins, notes that "there is a methodological problem" with relying on memory because the perpetrators that come to mind are at "the successful end of the distribution."[3] William Gamson adds that "we are likely to encounter a heavy bias toward success" because failures are harder to recall.[4] As Winston Churchill quipped, "History is written by the victors." Failures are never as newsworthy. My phone was off the hook when ISIS took Mosul. But I heard crickets when the group fell on its sword three bloody years later. This wasn't just my imagination. The *Washington Post* asked in December 2017: "Islamic State's caliphate has been toppled in Iraq and Syria. Why isn't anyone celebrating?"[5] Not only do assessments of terrorist effectiveness suffer from what methodologists call a "selection issue," but people are quick to declare the tactic a success even when the political effects weren't due to terrorizing civilians, such as when ISIS waltzed into Mosul as its guardians against the sectarian Iraqi government, or established themselves in unpopulated areas in the desert. When people point to successful cases of terrorist coercion, they're almost always examples of guerrilla attacks against military targets, like the oft-invoked Hezbollah violence behind the U.S., French, and Israeli withdrawals from southern Lebanon. As we've seen, lumping together such guerrilla attacks with terrorism contributes to the false impression that the tactic pays even though scholars generally define it as civilian-oriented violence.[6]

Terrorists are exalted as Joker-like criminal masterminds regardless of the outcome. That's because the result of their behavior is invariably hailed as exactly what the terrorists wanted to achieve.[7] This realization dawned on me in the immediate aftermath of the 9/11 attacks. At the time, Al Qaeda wasn't well known even among senior foreign policy members of the Bush administration.[8] So, I was surprised to find that everyone around me was sure that Al Qaeda had attained its aims by striking the United States. A local soccer coach told me that having to cancel soccer practice in the paralyzing fear after 9/11 was precisely what Al Qaeda secretly hoped to achieve. A cab driver told me that Al Qaeda must be delighted to know how many of his colleagues were staying home from work. Others said the specific date of 9/11 was brilliantly chosen to maximize the terror. "Those are the three digits that one dials here in America to call Emergency," one guy explained. "The terrorists wanted to plant the seeds of subliminal fear every time one calls 911...one would subconsciously remember 9/11 and thereby feel afraid."[9] But for the life of me, I couldn't recall anything in Bin Laden's fatwas about soccer practice, cab driving, or 911 operators. In the resulting panic weeks after the attacks, President Bush concluded, "These acts of mass murder were intended to frighten our nation into chaos."[10] With Americans hesitant to fly after the four planes were hijacked, he asserted, "They [the terrorists] want us to stop flying."[11] The toppling of the World Trade Center and the economic contraction that followed revealed that "the terrorists wanted our economy to stop." With American civil liberties restricted in the wake of the attacks, he proclaimed that Al Qaeda's goals, inter alia, were to curtail "our freedom of religion, our freedom of speech, our freedom to vote and assemble and disagree with each other."[12] Given that Al Qaeda has been largely mute on these views, it's hard to imagine Bush ascribing them to the terrorists had Americans not been frightened for their safety, hesitant to fly, and worried about their political and economic future in the wake of the 9/11 attacks.

Predictably, the result of ISIS behavior has also been portrayed as exactly what the group wants. When its evil atrocities became international news in the summer of 2014, commentators warned that "calling ISIS evil is what the terrorists want."[13] When a pair of brothers shot dead Charlie Hebdo cartoonists in January 2015, the *Washington Post* deemed the operation a success because "The terrorists wanted to kill people who make drawings for a living."[14] In November 2015, suicide bombers struck the Stade de France, terrorizing spectators watching *Les Bleus*. Analysts proclaimed that this effect was part of the master plan because "The message ... that the terrorists wanted to send is that nowhere is safe, even a stadium where the national team was playing."[15] Belgium responded to the Paris attacks by shutting down its public transportation. Right on cue, journalists said the closures were what "the terrorists wanted to achieve, and it seems that they have succeeded."[16] When the Paris attacks mobilized the world to crush ISIS, French President Francois Hollande remarked this is "precisely what ISIS intended" as "the deeper the West becomes involved in military action in the Middle East, the closer ISIS comes to its goal."[17] After ISIS was reduced to claiming credit for a botched lone-wolf bombing attack on the Tube that failed to detonate properly, the *New York Times* pontificated, "From ISIS' standpoint, this is a success" because the "purpose of terror is to terrorize" even with seemingly trivial attacks.[18] It's no wonder the media are constantly talking up terrorists as masterminds. Whatever happens to be the consequence is allegedly what the terrorists want.

Research in psychology sheds light on this curious phenomenon. In the 1960s and 1970s, the social psychologist, Edward Jones, identified a cognitive bias called correspondent inference theory. The theory builds on the foundational work of Fritz Heider, the father of attributional theory, to elucidate the cognitive process by which an observer infers the motives of an actor. Heider saw individuals as "naïve psychologists" driven by a practical concern—a need to simplify and comprehend the motives of others. Heider noted that humans don't have direct access into the minds of others, so we process information by applying inferential rules. In laboratory experiments, he found that people tend to attribute the behavior of others to inherent characteristics of their personality—or dispositions—rather than to external or situational factors. Correspondent inference theory resolves a crucial question that Heider left unanswered: How does an observer infer the motives of an actor based on his or her behavior? Jones showed that observers tend to interpret an actor's objective in terms of the consequence of the action. He offered the following simple example to illustrate the observer's assumption of similarity between the effect and objective of an actor: A boy notices his mother shut the door, and the room becomes less noisy, the correspondent inference is that she wanted to silence the racket coming from outside. If the room had instead become warmer, he would have likely concluded that she closed the door to get rid of the draft coming from outside. The essential point is what Jones called the "attribute-effect linkage," whereby the objectives of the actor are presumed to be encoded in the observable outcome of the behavior.[19] This is why soccer coaches, cab drivers, and everyone else in the world constantly say the terrorists are getting what they want. The effect of their actions is understood as the intent. So, terrorism is always seen as working regardless of the impact.

Even contradictory government responses are proffered as what terrorists want. When Bush responded to 9/11 by occupying Iraq in 2003, millions of Americans insisted that doing so was what Al Qaeda desired. When Spain had the opposite reaction after the 2004 Madrid train bombings by ending its occupation of Iraq, Americans said that's what Al Qaeda hoped for all along. Before the Paris attacks, the conventional wisdom held that ISIS wields violence to achieve a caliphate unadulterated by Western interference. After all, the leader was known to supporters as the "caliph." They were known to him as "soldiers of the caliphate." And ISIS members—both high and low—literally begged the U.S. and its allies not to intervene militarily against the caliphate.[20] But when ISIS attacks spurred an organization-crippling counterterrorism reprisal in Iraq and Syria, we were told that actually the Islamic State wants to provoke military interference in order to showcase Western brutality. Before the Paris attacks, we were told that France is a juicy target for the Islamic State because of the failure to integrate its Muslim population. After the attacks, we were told that the Islamic State actually wants France and the rest of the world to become even more xenophobic against Muslims.[21] Such mental gymnastics are ipso facto evidence of what the philosopher of science Imre Lakatos called a "degenerative research program." Research is degenerative when discrepant results are retroactively cast as consistent with expectations to keep the argument afloat.[22] Lakatos recognized that people are loath to ditch their theory. A telltale sign of intellectual bankruptcy is when they bend over backwards to sustain it by counting even contradictory outcomes as additional proof of its veracity.

Mea culpas for misdiagnosing ISIS have therefore been in short supply even after the caliphate imploded due to strategic blindness. Rather than retracting their repeated assertions of its strategic brilliance, pundits just moved the goalposts to redefine ISIS success in different, even opposite terms based on the latest consequences. Initially, there was widespread agreement that the goal of ISIS was to achieve a caliphate. Hassan Hassan, for example, was explicit that the best metric for evaluating ISIS is its territorial control.[23] Ticking off early territorial gains in Iraq and Syria, he said ISIS was manifestly successful because "it provides services, it has a military presence, it speaks as a state."[24] When ISIS proceeded to lose the caliphate, Hassan had a Lakatosian moment. The presumed importance of territory, he backpedaled, isn't what ISIS "really thinks." Contrary to what Baghdadi, his followers, and Hassan himself used to say, "Territorial control is not the main metric to judge ISIS' success."[25] After it got routed in Mosul and dispersed into the desert, for example, he insisted that ISIS' success remains undiminished because "the terrorist group views desert locations as being just as important as the big cities."[26] Similarly, Shadi Hamid used to exalt ISIS as strategic geniuses for the apparent success of the caliphate until it proved illusory. "They're not your terrorists of early to mid 2000s that were blowing things up and just killing innocent civilians without any kind of vision of how to build something," he assured us in 2014. Rather, "They actually control and hold territory."[27] When the caliphate began to implode a couple years later, he conceded: "The Islamic state may eventually be defeated. But their legacy will be long-lasting. ISIS has set the gold standard for extremist groups."[28] Suddenly, the measure of success wasn't an enduring caliphate as he had incorrectly prophesized. But the legacy of its politically futile head-chopping days. Charlie Winter of the International Centre for the Study of Radicalisation and Political Violence also shifted the goalposts to maintain his protean, unfalsifiable thesis about the strategic genius of ISIS leaders. Like other analysts, Winter originally said that the purpose of ISIS propaganda was "to show that the Islamic State is going to be around for a while."[29] In countless media interviews, he said Islamic State propaganda is "all geared around the caliphate cause and does have a purpose...to securitize and stabilize the area that they've created a level of hegemony over."[30] When the propaganda backfired on the caliphate, however, he said this must have been the master plan all along. "What if this 'catastrophe' is what ISIS wants?" he asked in *The Atlantic* magazine. "What if, more than anything else including territory, the group just wants...to be the ideological hegemon of global jihadism? In this pursuit, the realization of ideological aspirations is far more important than the permanent administration of any piece of land... Viewed through this lens, ISIS's most counter-intuitive acts become intuitive, if not ingenious, parts of a narrative-led strategy, one that prioritizes conceptual longevity over anything else."[31] In these ways, terrorists are continuously hailed as ingenious success stories whether they accomplish their stated mission or not, as Jones and Lakatos would appreciate.

The terrorists themselves are the first to move the goalposts so as not to seem like losers. When the shrinking of the caliphate became incontrovertible in May 2016, for instance, ISIS spokesman, Abu Muhammad Al Adnani, released a statement pundits would quote ad infinitum: "Do you think,

O America, that defeat is the loss of a city or a land? Were we defeated when we lost cities in Iraq and were left in the desert? Will we be defeated and you will be victorious if you took Mosul or Sirte or Raqqa or all the cities, and we returned to where we were in the first stage? No, defeat is the loss of willpower and the desire to fight."[32] In this way, ISIS leaders lowered the bar of success from its ambitious goal of a caliphate to nothing but grit. Such dissembling became routine as the caliphate charade was exposed. In July 2017, for instance, ISIS propaganda went from warning the U.S. not to destroy its caliphate to insisting that the organization-crushing air campaign "only added to our faith and courage."[33] Perhaps the silliest example of goalpost-moving was over Dabiq. ISIS had long promised that a showdown there would defeat the crusaders in an epic apocalyptic battle. So foundational was this lore in ISIS propaganda that the group named its main online magazine *Dabiq*. The actual battle was epic, but not as ISIS had prophesized. The group lost the village in a few hours. To soften the blow, ISIS trivialized the trouncing and promised the real battle would come at some undisclosed time.[34] Of course, pundits were happy to parrot back the terrorist talking points to preserve the mastermind fantasy and avoid recalling years of faulty analysis.[35]

How else could terrorist leaders gain and maintain members without cultivating false hope? As Mark Lichbach notes, "Dissident entrepreneurs attempt to justify sacrifice and sustain hope by convincing people that . . . history is on our side and that defeat is temporary, victory inevitable."[36] In *The Rebel Dilemma*, he notes that leaders have "an eye toward increasing the perceived probability of winning," make "unqualified predictions of eventual victory," and "inflate revolutionary expectations by maintaining that they have had victories, continue to win, and will eventually succeed."[37] The victory illusion is based on the efficacy illusion, the notion that members are contributing to the goal contrary to all appearances.[38] Walter Laqueur and Yonah Alexander also remark, "Revolutionaries aim to create an impression of power, invincibility and effectiveness, representing themselves as a force which must inevitably assume supremacy."[39] Stephen Walt has made a related observation: "Unless potential supporters believe their sacrifices will eventually bear fruit, a revolutionary movement will not get very far. Revolutionary ideologies are thus inherently optimistic: they portray victory as inevitable despite what may appear overwhelming odds."[40] Richard English points out that the leaders thus "exaggerate the possibility of terrorism working."[41] Bin Laden, for example, changed the 9/11 narrative so as not to seem like a fool. As John Mueller tells it:

> Initially, there was panic in Al Qaeda at the unexpected ferocity of the American response. Then bin Laden reformulated his theory after it was blown to shreds when the United States and its allies not only forced Al Qaeda out of its base in

Afghanistan and captured or killed many of its main people but also toppled the country's accommodating Taliban regime. In a videotaped message in 2004, bin Laden mockingly asserted that is 'easy for us to provoke and bait... All that we have to do is to send two mujahidin... to raise a piece of cloth on which is written Al Qaeda in order to make the generals race there to cause America to suffer human, economic, and political losses...' But that is more nearly a convenient rationalization than a fair representation of his goals when he had planned the attack."[42]

Donatella Della Porta likewise describes how the Italian and German underground groups in the 1970s "cultivated an overly optimistic view of the success of the armed struggle" with dialectical gibberish about how "the worse the better" for the "the coming revolution."[43] Even successful groups like Hezbollah exaggerate their effectiveness. Thanassis Cambanis notes: "Hezbollah decided from its inception that it could always present a victorious face to its public as long as it carefully defined victory and adroitly redefined it when circumstances changed... Sayyed Nasrallah in a single speech could frame any confrontation to Hezbollah's advantage, spotlighting success where he could find a trace and resiliency where he could not... Even if Israel ultimately took over all of Lebanon, by this logic, Hezbollah would have won, first by frustrating Israel's original military plan, and second, by exposing the evil nature of Israel. It was a brilliant strategy because in almost every conceivable case it gave Nasrallah license to define any real-world outcome—no matter how disastrous—as a victory for Hezbollah and its people, and thus to entrench his political power."[44] For example, Nasrallah publicly vowed to avenge the Mossad assassination of Imad Mughniyah. Upon failing to retaliate against Israel for killing him with a car bomb in 2008, Hezbollah said the worst vengeance was keeping the threat of an impending attack hanging over Israel's head. Matthew Levitt describes the spin, "The absence of an attack, then, was not a sign of failure but part of a master plan."[45] Like other groups, Hezbollah also inflates the number of enemy casualties while underreporting its own losses.[46] The overselling of terrorist success is brilliantly captured in the film *Four Lions*. In this 2010 British black comedy, bumbling terrorists pretend that accidentally blowing up sheep was a watershed strike on the infidels' critical food infrastructure.[47] By repeating their talking points, pundits thus exaggerate public perceptions of terrorist skill and accomplishment.

Opinion-makers weaponize the prospect of terrorist success to advance their policy preferences. When politicians and commentators want to erode support for an unwanted political position, they just say it plays into the hands of the terrorists. Few people want to help terrorists, so saying they'll benefit from the policy is an effective way to counter it. In June 2017, for example, the leader of the UK Independence Party didn't want to compromise his run in the upcoming general election after ISIS attacked the London Bridge. So, he argued that heeding the post-attack moratorium on campaigning is "what the extremists would want."[48] When Senator Lindsay Graham didn't like a budgetary adjustment to the State Department, he said ISIS "will be celebrating" if the proposal gets approved.[49] In the presidential campaign, Hillary Clinton suggested that electing Trump would "give ISIS what it wants."[50] MSNBC repeated that Trump is "the president ISIS wants."[51] The New York Times warned there's "no question ISIS wants Trump as president."[52] New York Magazine claimed, "ISIS seems to be a fan of Trump . . . the single most brutal and successful radical Islamist terrorist group in history is actually rooting for a particular candidate in this American election."[53] CNN likewise asserted, "A Trump White House would be a real coup for ISIS."[54] When this argument failed to carry the day, critics tried to sink his policies by saying they're exactly what ISIS hopes will happen. As president, Trump issued Executive Order 13769 to temporarily ban nationals of seven Muslim-majority countries from entering the United States for at least the next ninety days. His political opponents derided these restrictions as not only Islamophobic, but a major victory for ISIS. For example, CNN.Com posted pieces entitled "ISIS is celebrating Trump's immigration ban"[55] and "Trump ban is boon for ISIS recruitment."[56] Many other outlets leveled related allegations, like the Huffington Post, which claimed that limiting refugees into the U.S. "will be playing directly into the hands of extremists."[57] These talking points were driven by the desire to oppose Trump rather than objective assessments of the ISIS threat. The narrative fell apart when reporters realized that ISIS wasn't talking about Trump at all—never mind his policies. A perplexed New York Times reporter confessed, "I don't know what to make of ISIS's silence on Trump."[58] A Slate reporter admitted that ISIS members seem to be "the only people on Earth who are not talking about Trump apparently."[59] Politico remarked: "The consensus among liberals, prominent terrorism experts and even some conservatives is that the jihadists are enthused, in a gleeful, hand-rubbing sort of way, by his presidency and that they warmly welcome the self-inflicted wound of the executive order on refugees as a propaganda victory . . . The reality, however, is that ISIS has remained conspicuously silent on Trump's presidency, let alone any of his policies . . . Despite repeated claims that the travel ban is a propaganda victory for ISIS and that the group will use it to attract new recruits, the ban has yet to be mentioned by ISIS leaders."[60] To maintain the mastermind myth, though, pundits then claimed that the silence was strategic because it allowed Trump to tarnish his own brand, which they simultaneously said worked in the group's favor.[61] In effect, opinion-makers weaponize the prospect of terrorism working to serve their own policy preferences, stoking the unfalsifiable public perception of the terrorist mastermind.

Meanwhile, authorities depict the terrorist enemy as supremely clever so they don't get egg on their faces in the event of a future attack. In the aftermath of 9/11, terrorism watchers were sure a second shoe would drop. And yet, we kept waiting for another major strike on the U.S. homeland to materialize. Among those in the know, the belief quietly spread that perhaps Al Qaeda doesn't have the chops for a commensurate attack in the tougher post-9/11 counterterrorism climate. But nobody wanted to admit this publicly lest an attack got through. Michael Sheehan, New York City's deputy director for counterterrorism, told his bosses, Raymond Kelly and David Cohen, that "Al Qaeda was simply not very good . . . under the withering heat of the post-9/11 environment, they were simply not getting it done." Sheehan recounts the reaction of colleagues: "They were all taken aback. It was not so much that they disagreed . . . They all understood only too well the way the public and politicians would react if headlines started to read 'Commissioner Disses Qaeda.'"[62] Not only does downplaying the threat risk backfiring if it happens, but threat deflation risks drying up counterterrorism funding. And for terrorism commentators, less fear means fewer media hits.[63] Given these concerns, "Fearmongering by officials and by the media is politically (and economically) understandable," as Mueller has noted.[64]

Several global trends are also conducive to a rise in terrorism beyond the interminable characterizations of its effectiveness. For starters, the proliferation of drone technology means that more decapitation strikes against militant groups will attenuate their leaderships, empowering lower-level members with weaker restraint toward civilians. Similarly, I foresee an upward trend in lone-wolf attacks as counterterrorism measures make it harder to link up with militant groups and improved communications make it easier to self-radicalize. Without strong leaders to restrain them, lone wolves will keep preying on civilian targets.[65] But by hardening targets, governments are inadvertently incentivizing attacks on softer ones. Researchers have uncovered a "substitution effect" in which attackers gravitate to softer targets when higher-value ones have been hardened.[66] In the nineteenth century, increased security for kings and tsars gave way to attacks on less consequential figures.[67] Over the past few decades, perpetrators have directed more attacks against business and other private targets as official ones have become harder to strike.[68] And as the quality of government countermeasures increases, the quality of perpetrators will decline. Being a terrorist today is a riskier enterprise than ever before. Consider that in the 1960s terrorists were less likely to become a casualty than those who participated in rioting.[69] RAND found in a sample of 127 hijack attempts between 1968 and 1974 that the perpetrators had less than a 10 percent chance of being killed or imprisoned.[70] And sentences used to be light. At least in the West, no European, North American, Japanese, or Middle Eastern terrorist of the 1960s or 1970s was executed.[71] The average sentence awarded to terrorists was just eighteen months.[72] There was always a good chance of being released before serving out a term.[73] According to the U.S. Secretary of State and Coordinator for Combating Terrorism, between 1971 and 1974 less than 50 percent of captured international terrorists actually served the full prison term.[74] The heightened risk these days dissuades many prospective terrorists. Remaining ones are more risk-acceptant, impulsive, even unstable. Tougher counterterrorism responses also risk punishing innocent people, spurring them to radicalize, and attack out of vengeance regardless of the fallout. Terrorism will thus continue despite the poor political return.

Counterterrorism Implications

The international community can help to reduce the incentive and thus incidence of terrorism. For starters, opinion-makers need to stop describing terrorists as extreme Einsteins. Beyond its inaccuracy, this characterization is dangerous. It inspires people to attack civilians for the hagiographies to follow. And it promotes false hope by interfering with the natural learning process that terrorism doesn't actually pay politically. As we've seen, target countries are reluctant to appease terrorism. International bodies like the United Nations must think twice about foisting political settlements on target countries to avoid spurring aggrieved groups in other theaters to attack civilians. If anything, the international community should reward aggrieved parties for nonviolent protest. And the media should tout their underappreciated successes.[75]

In addition to digging in their political heels in the face of terrorism, governments should carefully consider the role of each militant leader before taking him out. There's a tendency to conclude that because militant groups are extreme, their leaders are the most radical, so they should be neutralized. As we've seen, though, removing senior members in most militant groups leads to indiscriminate rampages like the bull-less elephants from Kruger National Park. We must be especially careful about removing strong leaders who oppose civilian attacks because the next generation of the group will become even more extreme. Similarly, governments must resist overreacting to terrorism with indiscriminate punishments that will spur terrorist recruitment out of vengeance. Indiscriminate violence is counterproductive for state and non-state actors alike. Avoiding these missteps will help militant leaders— at least the smart ones—to restrain the rank-and-file from lashing out in feckless fits of rage.

When militant leaders are stupid enough to broadcast their massacres let's not stand in their way. Sometimes the best strategy against terrorists is doing exactly what they want. Because ISIS was supposedly benefiting from its snuff videos, pundits urged social media platforms like Twitter to close down the gore-packed propaganda accounts.[76] Yet our response to terrorist propaganda must be research-based. The metric for success shouldn't be the number of social media accounts we've shut down, but the extent to which we've heightened the audience costs by sullying the image of the group. Governments and the private sector should therefore highlight the evil face of the group to expedite the worldwide turn against it. Journalists will continue to focus on every errant example of crazed recruits attracted to the carnage. But the decency of most people in the world will in the aggregate push the attrition rate from the violence above the recruitment rate to the detriment of the terrorists.

Finally, let's get *Rules for Rebels* into the hands of militant leaders. Following the rules will help them achieve their goals. But this will also help ours by keeping us safe.

Notes

1. Tversky and Kahneman 1973.
2. Dershowitz 2002, 7.
3. Collins 2001, 34.
4. Gamson 1975, 22.
5. El-Ghobashy, Salim and Loveluck, December 5, 2017.
6. Abrahms, November 8, 2017.
7. Abrahms, November 20, 2015.
8. Scheuer 2004.
9. Earthmann, June 23, 2014.
10. Bush, November 17, 2001.
11. Ibid.
12. Bush, September 23, 2001.
13. Colmes, August 25, 2014.
14. "Why Would Terrorists Kill Cartoonists?" January 8, 2015.
15. Moritz 2015.
16. "Security Over Freedom," November 25, 2015.
17. Atran and Hamid, November 16, 2015.
18. Callimachi, September 15, 2017.
19. Jones and Davis 1965; Jones and McGillis 1976.
20. Stern and Berger 2015, 159.
21. Nance 2016, 309.
22. See Vasquez 1997.
23. Hassan, August 9, 2014.
24. Schreck and Karam, September 2, 2014.
25. Hassan, July 2, 2017.
26. Bob, November 17, 2016.
27. "How Is the Islamic State Different?" August 21, 2014.
28. Malley, July 6, 2016.
29. Sharkov December 10, 2014.
30. Alfred, October 12, 2015.
31. Winter, July 3, 2017.

32. Hassan, July 2, 2017.
33. Hubbard and Schmitt, July 8, 2017.
34. Barnard, October 19, 2016.
35. Gerges 2016; Hassan, July 3, 2017; Ward, July 6, 2017.
36. Lichbach 1998, 23.
37. Ibid. 74–5.
38. Ibid. 91.
39. Laqueur and Alexander 1978, 237.
40. Walt 1992, 337.
41. English 2016, 6.
42. Mueller and Stewart 2016, 120.
43. Della Porta 2006, 175.
44. Cambanis 2011, 82.
45. Levitt 2013, 4.
46. Gleis and Berti 2012, 90.
47. Morris 2010.
48. "London Attack," June 4, 2017.
49. Savage and Schmitt, March 12, 2017.
50. Roberts, September 19, 2016.
51. "Donald Trump: The President ISIS Wants?" May 24, 2016.
52. Ibid.
53. Anderson, October 26, 2016.
54. Winter, November 3, 2016.
55. Burke, January 31, 2017.
56. Mackintosh, January 31, 2017.
57. Cooke, November 16, 2015.
58. Chotiner, July 12, 2016.
59. Ibid.
60. Cottee, March 26, 2017.
61. Heneghan, March 17, 2017.
62. Mueller and Stewart 2016, 50.
63. Lula 2013.
64. Mueller and Stewart 2016, 50.
65. See Becker 2014.
66. See Enders and Sandler 2004.
67. Lichbach 1998, 359.
68. Brandt and Sandler 2009; Dau, Moore, and Abrahms 2018.
69. Gurr 1979, 38.
70. Jenkins et al. 1977, 92.
71. Laqueur 1978, 92.
72. Wilkinson 1979, 115.
73. Laqueur 1978, 92.
74. Wilkinson 1979, 115.
75. Chenoweth and Stephan 2008.
76. See, for example, Berger and Morgan 2015.

Appendix of Tables

Table A.1. Determinants of FTO Campaign Success

	Logit	Ordered logit	Ordered logit Without AQ	Ordered logit without Iraq, Afghanistan
	(1)	(2)	(3)	(4)
Civilian target (1/0)	−7.33**	−4.06***	−3.19***	−4.20***
	(2.39)	(0.76)	(0.85)	(0.89)
Maximalist objective (1/0)	−6.82***	−1.69***	−1.89**	−1.85**
	(2.01)	(0.52)	(0.64)	(0.64)
FTO capability:				
Peak membership (100s)	−0.012*	−0.009***	−0.003	−0.01***
	(0.005)	(0.002)	(0.004)	(0.003)
FTO lifespan (in years)	0.004	0.01	0.04	0.06*
	(0.04)	(0.05)	(0.03)	(0.03)
External support (1/0)	0.27	0.70	1.04	1.03
	(1.35)	(0.52)	(0.61)	(0.68)
Suicide Tactics (1/0)	0.09	0.30	0.93	0.88
	(0.72)	(0.53)	(0.80)	(0.83)
Target capability:				
Polity IV score	−0.26	0.07	0.03	0.04
	(0.13)	(0.07)	(0.09)	(0.10)
Per capita GDP (log)	1.41	−0.03	1.17	0.63
	(1.53)	(0.51)	(0.75)	(0.78)
Material capabilities (log)	0.54	0.33	0.03	0.25
	(0.39)	(0.24)	(0.39)	(0.44)
Population (log)	−0.36	−0.16	0.47	0.31
	(0.50)	(0.26)	(0.34)	(0.36)
Constant/Cutpoints	−17.17	−18.87	−73.01	−111.76
	(62.75)	−17.43	−70.02	−108.46
		−14.44	−68.94	−106.91
Pseudo R^2	0.67	0.29	0.29	0.35
N	117	117	70	73

Notes: Logit (Column 1) and Ordered Logit (Columns 2, 3, 4) estimations with standard errors in parentheses. *$p < 0.05$, **$p < 0.01$, ***$p < 0.001$. In the logit model, complete failures and near failures are coded 0, and partial successes and complete successes are coded 1. The four dependent variable outcomes in the ordered logit models are complete failure, near failure, partial success, and complete success.

Table A.2. Marginal Effects of Statistically Significant Variables

	Marginal Effect (Change of Probabilities)			
	Complete Failure	Near Failure	Partial Success	Complete Success
Mil → Civ Target	0.77	−0.12	−0.55	−0.10
Lim → Max Objective	0.40	−0.16	−0.22	−0.02

Note: Probabilities are calculated holding all other independent variables at their means.

Table A.3. Determinants of Bargaining Success

Explanatory Variables	Model 1	Model 2	Model 3	Model 4
Constant	−3.444	−2.795	−2.786	−2.789
	(0.335)***	(0.233)***	(0.233)***	(0.228)***
COW Capability Index	−0.3956	−0.176	−0.526	−0.562
	(6.057)	(6.032)	(6.143)	(6.132)
Polity 2 Score	−0.032	−0.033	−0.031	−0.032
	(0.017)·	(0.0169)·	(0.018)·	(0.018)·
Group Size	0.00002	0.00002	0.00002	0.00002
	(0.00005)	(0.00005)	(0.00004)	(0.00005)
Hostages	0.006	0.006	0.006	0.006
	(0.003)·	(0.003)·	(0.003)·	(0.003)·
Financial Ransom	1.0567	1.063	1.05	1.0358
	(0.383)*	(0.383)*	(0.383)*	(0.395)*
Prisoner Release	−0.217	−0.242	−0.205	−0.193
	(0.531)	(0.532)	(0.533)	(0.519)
Safe Haven Demand	0.27	0.288	0.27	0.26
	(0.727)	(0.725)	(0.731)	(0.733)
Safe Passage Demand	1.378	1.413	1.412	1.437
	(1.132)	(1.32)	(1.148)	(1.156)
Robin Hood Ransom Demand	1.328	1.507	1.271	1.31
	(7.35)	(7.479)	(7.313)	(7.333)
Number of Demands	1.029	1.018	1.03	1.031
	(0.16)***	(0.159)***	(0.16)***	(0.161)***
Non-Government Target	1.036	1.049	1.035	1.033
	(0.294)**	(0.293)**	(0.295)**	(0.293)**
No Physical Harm	0.6558			
	(0.282)*			
Civilians Harmed		−0.6601	−0.634	−0.687
		(0.32)*	(0.32)*	(0.33)*
Officials Harmed			−0.6813	−0.773
			(0.5)	(0.517)
Negotiation Harm				0.524
				(0.718)
Observations	1075	1075	1075	1075

Note: Each column reports the estimated coefficients of a separate logit regression in which the dependent variable is negotiation success, which equals 1 if the group received any demands and 0 otherwise. Robust standard errors appear in parentheses. '***'significant at 0.001; '**'significant at 0.01; '*'significant at 0.05; '·'significant 0.1.

Table A.4. Estimated Effect of Expected First Differences on Negotiation Success

	Model 1		Model 2	
	Δ Hostages	Δ Terrorism	Δ Hostages	Δ Terrorism
Δ Prob Success %	0.0677	−6.74	0.066	−6.89
	(0.0023–0.224)	(−22.24––0.375)	(0.0006–0.22)	(−23.08––0.25)

Note: Each column reports the estimated effect of first differences on the probability of negotiation success (%). The parentheses contain a 95% confidence interval for each expected first difference. The change in the number of hostages is 1 to 2 hostage captured. The terrorism variables change from 0 (not present) to 1 (present). The tactical variable of terrorism in Model 1 includes incidents in which civilians and/or government officials are physically harmed whereas terrorism in Model 2 refers only to incidents in which civilians are harmed.

Table A.5. Differences between Painful Treatment and Painless Control (Vignette 1)

	Question 1	Question 2 (y | Opinion)	Question 3	Question 4	Question 5
Painful Treatment	0.656	3.306	0.738	0.804	0.697
Painless Control	0.381	4.616	0.501	0.47	0.481
Difference	0.274	−1.31	0.236	0.334	0.216
	(0.207–0.342)	(−1.596––1.024)	(0.171–0.302)	(0.271–0.397)	(0.149–0.283)

Note: Vignette 1 refers to the scenario in which the hostage-takers demand prisoners in exchange for permanently demobilizing. Each column reports the average of the painful treatment and painless control for each survey question. Question 2 reports the average given that respondents had an opinion on a 7-point ordinal scale. 95 percent confidence intervals are in parentheses for each two-tailed difference of means test.

Table A.6. Differences between Painful Treatment and Painless Control (Vignette 2)

	Question 1	Question 2 (y | Opinion)	Question 3	Question 4	Question 5
Painful Treatment	0.693	3.296	0.744	0.798	0.687
Painless Control	0.294	5.031	0.39	0.418	0.425
Difference	0.3998	−1.7354	0.354	0.38	0.262
	(0.336–0.464)	(−2.018––1.452)	(0.29–0.419)	(0.318–0.443)	(0.195–0.329)

Note: Vignette 2 refers to the scenario in which the hostage-takers demand money in exchange for permanently demobilizing. Each column reports the average of the painful treatment and painless control for each survey question. Question 2 reports the average given that respondents had an opinion on a 7-point ordinal scale. 95 percent confidence intervals are in parentheses for each two-tailed difference of means test.

Table A.7. Audience Costs of Terrorism

Variables	Repression	Violence
Civilian Attack	0.70*	1.43**
	(0.32)	(0.45)
Intercept	−3.67***	−7.59***
	(0.41)	(0.72)
Wald Chi Squared	4.71*	9.93**
Log Likelihood	−501.90	−152.54
Terrorist organizations (groups)	112	112
N	1,677	1,677

Note: Standard errors appear in parentheses.
Significance levels (2-tailed): †p<0.10, *p<0.05, **p<0.01, ***p<0.001.

Table A.8. Determinants of Civilian Targeting with MAROB Data

	Model 1	Model 2	Model 3
	β/(SE)	β/(SE)	β/(SE)
Violence Abroad		1.784***	
		(0.210)	
Cross-border Insurgency			1.958***
			(0.355)
Organizational Properties			
Organizational Age	−0.044***	−0.026+	−0.035**
	(0.013)	(0.014)	(0.014)
Organizational Size	0.594***	0.667***	0.753***
	(0.161)	(0.181)	(0.166)
Territorial Control	0.666*	0.855**	0.852**
	(0.270)	(0.293)	(0.275)
Popular Support	−0.295	−0.521	−0.623+
	(0.342)	(0.340)	(0.341)
Propaganda	0.693***	0.554**	0.651***
	(0.168)	(0.189)	(0.173)
Religious	0.216	0.701+	0.314
	(0.314)	(0.378)	(0.351)
Ethno-nationalist	1.024*	1.052*	1.092**
	(0.437)	(0.479)	(0.422)
State Properties			
Population Density	−0.000	−0.000	−0.000*
	(0.000)	(0.000)	(0.000)
Education	−0.001*	−0.002**	−0.002***
	(0.001)	(0.001)	(0.001)
GDPPC	0.000+	0.000***	0.000**
	(0.000)	(0.000)	(0.000)
Regime	−2.319***	−2.355***	−2.516***
	(0.403)	(0.454)	(0.398)
Effective Executive	1.606***	1.346***	1.629***
	(0.294)	(0.331)	(0.302)
Conflict Index	−0.000*	−0.000**	−0.000**
	(0.000)	(0.000)	(0.000)
N	668	655	665

+ <0.10, *p<0.05, **p<0.01, ***p<0.001

Note: Estimates are maximum likelihood coefficients from logit equations with the organization year as the unit of analysis. Robust standard errors are in parentheses.

Table A.9. Determinants of Civilian Targeting with BAAD Data

	Model 1: Percent Civilian	Model 2: Percent Civilian	Model 3: Percent Civilian (No AQ)
Affiliates	0.115	0.130	0.136
	(0.051)	(0.038)	(0.036)
Religious ideology		0.354	0.354
		(0.000)	(0.000)
Ethno-nationalist ideology		0.253	0.251
		(0.003)	(0.003)
Ethno-nationalist & religious		0.183	0.187
		(0.053)	(0.051)
Leftist ideology		0.042	0.042
		(0.645)	(0.649)
State sponsorship		0.036	0.038
		(0.699)	(0.685)
Size (ordinal)		−0.006	−0.003
		(0.869)	(0.943)
Control of territory		0.123	0.123
		(0.119)	(0.121)
Organizational connections		−0.004	−0.005
		(0.730)	(0.670)
Count, fatalities		−0.000	−0.000
		(0.807)	(0.645)
Intercept	0.434	0.234	0.235
	(0.000)	(0.001)	(0.001)
R^2	0.016	0.140	0.140
Adjusted R^2	0.012	0.095	0.095
N	238	202	201

Note: p-values (2-tailed) in parentheses.

Table A.10. Learning to Win

	Short-term	Long-term
Civilian attack	$p–c_{cs}$	$p–c_{cl}$
Military attack	$p–c_{ms}$	$p–c_{ml}$
Civilian attack	2–1	2–1
Military attack	2–3	2–0
Civilian attack	1	1
Military attack	−1	2

Note: p = payoffs; c = costs; c_{cs} = short-term civilian costs; c_{ms} = short-term military costs; c_{cl} = long-term civilian costs; c_{ml} = long-term military costs.

215

Table A.11. Video Properties and Selectivity

	Dependent variable
	Selectivity
Subtitles	0.568
	(0.514)
Production quality	0.037
	(0.248)
Number of speakers	0.142
	(0.145)
Music	0.168
	(0.539)
Asymmetric attacks	1.097**
	(0.482)
Hostages	−0.511
	(0.580)
Anger	−1.572***
	(0.524)
Spokesman	0.307
	(0.520)
Official	−0.591
	(0.592)
Constant	−0.458
	(1.228)
Observations	116
Akaike Inf. Crit.	158.652

Note: *p < 0.1; **p < 0.05; ***p < 0.01.

Table A.12. Civilian Attacks

	Model 1	Model 2	Model 3	Model 4	Model 5	Model 6	Model 7
	β/(SE)	β/(SE)	β/(SE)	β/(SE)	β/(SE)	β/(SE)	β/(SE)
Organizational Structure	−0.473*	−0.511*	−0.438	−0.731	0.892	−0.142	−0.244
	(0.207)	(0.230)	(0.293)	(0.517)	(0.963)	(0.352)	(0.614)
Leader Sanction		16.214***	14.708***	15.676***	8.630***	15.735***	14.570***
		(0.426)	(0.534)	(0.722)	(1.692)	(0.653)	(1.429)
Structure*Sanction		−13.362***	−12.340***	−14.135***	−6.907***	−13.683***	−12.960***
		(0.527)	(0.605)	(0.795)	(1.742)	(0.710)	(1.529)
Organizational Properties							
Organizational Age			−0.026**	−0.062***	0.604	−0.026**	1.917*
			(0.009)	(0.016)	(1.145)	(0.009)	(0.910)
Organizational Size			−0.045	0.467*	0.217	0.053	−0.942*
			(0.121)	(0.222)	(0.980)	(0.142)	(0.473)
Territorial Control			0.119	0.586*	−2.034*	0.454+	0.092
			(0.179)	(0.294)	(0.910)	(0.232)	(0.377)
Popular Support			−0.013	0.136	0.837	0.055	0.814
			(0.263)	(0.392)	(2.004)	(0.275)	(1.304)
Propaganda			0.880***	0.916***	2.417***	0.825***	1.287**
			(0.155)	(0.204)	(0.495)	(0.168)	(0.441)
Religious			0.163	0.611	14.109	0.792**	23.808*
			(0.242)	(0.439)	(22.693)	(0.289)	(9.602)
Ethno-nationalist			−0.462*	0.827+	25.308**	0.567+	−32.952*
			(0.210)	(0.501)	(9.697)	(0.342)	(13.713)
State Properties							
Population Density				−0.000	0.004*	−0.000	0.001
				(0.000)	(0.002)	(0.000)	(0.000)
Primary & Secondary Education				−0.001*	−0.005**	−0.001**	−0.001
				(0.001)	(0.002)	(0.000)	(0.001)
GDPPC				0.000	0.001**	0.000	0.000
				(0.000)	(0.000)	(0.000)	(0.000)
Regime				−2.292***	−16.229***	−1.341***	−1.658*
				(0.429)	(3.692)	(0.258)	(0.777)
Effective Executive				2.000***	16.965***	1.342***	2.115**
				(0.362)	(3.709)	(0.233)	(0.799)
Conflict Index				−0.000	−0.000*	0.000	0.000
				(0.000)	(0.000)	(0.000)	(0.000)
N	998	997	910	587	407	1004	749

+ <0.10, *p<0.05, **p<0.01, ***p<0.001

Note: Estimates are maximum likelihood coefficients from logit equations with the organization year as the unit of analysis. Robust standard errors are in parentheses. All models include year-fixed effects.

Table A.13. Decapitation Strikes and Target Selection

	Model 1	Model 2	Model 3
	Civ. Targets/Attacks	Civilian Targets	Military Targets
	β/(SE)	β/(SE)	β/(SE)
Leaders Killed	0.275*	0.272***	−0.246
	(0.126)	(0.076)	(0.181)
Leaders Killed $_{t-1}$	0.043	−0.040	−0.500**
	(0.126)	(0.082)	(0.191)
Leaders Killed $_{t-2}$	−0.005	0.027	−0.776***
	(0.126)	(0.082)	(0.207)
Drone Strikes	0.209***	0.140***	−0.118
	(0.058)	(0.037)	(0.072)
Drone Strikes $_{t-1}$	0.150**	0.165***	0.001
	(0.057)	(0.036)	(0.066)
Drone Strikes $_{t-2}$	0.298***	0.181***	0.013
	(0.058)	(0.037)	(0.067)
N	2902	2902	2902

+ < 0.10, *$p < 0.05$, **$p < 0.01$, ***$p < 0.001$

Note: Model 1 is a fractional logit. Coefficients in Models 2 and 3 are derived from negative binominal models. All models have a daily unit of analysis and contain week fixed effects.

Table A.14. Mean Comparison of Decapitation Attempts

	Successful	Unsuccessful	Difference
	Afghanistan and Pakistan		
Proportion of Military Targets	0.1524	0.1968	−0.0444
Number of Attacks	7.9250	7.7600	0.1650
	Israel-West Bank-Gaza Strip		
Proportion of Military Targets	0.2182	0.3900	−0.1718*
Number of Attacks	0.8091	0.8000	0.0091

Table A.15. Decapitation Strikes on Civilian Attacks

Afghanistan and Pakistan

	Military Attacks	Number of Attacks
Success	−0.358* (0.186)	0.0170 (0.0676)
Bush	Reference	Reference
Obama	−0.199 (0.277)	−0.0769 (0.0663)
Osama	0.591* (0.311)	−0.103 (0.0854)
Malakand Accord	−0.281 (0.269)	0.0618 (0.134)
Constant	−1.546*** (0.269)	2.110*** (0.0592)
Offset	ln(Number of Attacks)	None
Observations	90	90

Israel-West Bank-Gaza Strip

	Military Attacks	Number of Attacks
Success	−0.589* (0.311)	−0.0206 (0.322)
Barak-Sharon	Reference	Reference
Sharon-9/11	−1.233*** (0.407)	−.918* (0.988)
9/12-ODS	−1.550*** (0.315)	2.665*** (1.017)
ODS-Roadmap	−18.00*** (0.556)	2.294** (1.041)
Roadmap-Ceasefire	−1.186*** (0.221)	2.118** (0.976)
Ceasefire	See Notes	−12.13*** (1.273)
Post-Ceasefire	−1.618*** (0.499)	1.931* (1.000)
Fatah	Reference	Reference
Hamas	0.330 (0.386)	−0.106 (0.243)
Islamic Jihad	0.410 (0.606)	−1.330*** (0.462)
Constant	0.259 (0.437)	−2.066** (1.003)
Offset	ln(Number of Attacks)	None
Observations	63	130

Notes: */**/*** indicate statistical significance at the 10%/5%/1% levels, robust standard errors in parentheses. Estimation based on a Poisson count model of the number and targets of attacks in the fourteen days after a decapitation attempt. The offset indicates what, if applicable, the dependent variable was conditioned on. If no attacks occurred after a targeted killing, that observation is automatically dropped from the estimation sample for the Military Attacks estimation. The Ceasefire dummy is dropped because none of the decapitation attempts in that period were followed by attacks.

Table A.16. Incidence Rate Ratio of Targeted Killing

	Afghanistan and Pakistan
Military Attacks	0.699*
	(0.130)
Number of Attacks	1.017
	(0.0688)
	Israel-West Bank-Gaza Strip
Military Attacks	0.555*
	(0.173)
Number of Attacks	0.980
	(0.315)

Notes: The cell entries indicate the factor by which a successful targeted killing affects the number of attacks or the proportion of military attacks.

Table A.17. Effect of Time on Tactics

	Seasonal Effects	Ramadan Effect	Third & Fourth	Fifth & Sixth
Afghanistan and Pakistan	−0.306*	−0.346*	−0.281	0.0637
	(0.166)	(0.197)	(0.175)	(0.194)
Israel-West Bank-Gaza Strip	−0.763**	−0.589*	0.328	0.587
	(0.331)	(0.314)	(0.551)	(0.398)

Notes: Dependent variable is target choice. Column 3 focuses on the third and fourth week after the decapitation attempt. Column 4 focuses on the fifth and sixth week.

Table A.18. Attack Claims by Target Selection

	Military/Government	Civilian
Not claimed	5,663	13,423
Claimed	1,115	1,758
Observations	6,778	15,181

Table A.19. Competing Claims by Target Selection

	Military/Government	Civilian
No competing claims	192	399
Competing claims	34	40
Observations	226	439

Table A.20. Determinants of Credit Claims

	Model 1	Model 2	Model 3	Model 4
Civilian Target	0.40***	0.39***	−0.27**	−0.24ᵃ
	(0.08)	(0.09)	(0.13)	(0.13)
Number of Fatalities	0.01***	0.01***	0.01	−0.01
	(0.01)	(0.01)	(0.01)	(0.01)
State Repression	−0.20	−0.19	−0.03	−0.06
	(0.13)	(0.15)	(0.16)	(0.16)
Assassination	−0.19	−0.21	0.06	−0.01
	(0.24)	(0.26)	(0.33)	(0.37)
Hostage Incident	0.62**	0.74**	0.18	0.20
	(0.28)	(0.31)	(0.22)	(0.19)
Armed Assault	−0.22	−0.23	−0.18	−0.28
	(0.15)	(0.17)	(0.17)	(0.18)
Multiparty Competition (GTD)	−0.39**	−	0.13	−
	(0.21)	−	(0.21)	−
Number of Groups	−	0.02	−	0.05***
	−	(0.01)	−	(0.02)
Organization Size	−	−	−0.92***	−0.75***
	−	−	(0.27)	(0.25)
Foreign State Sponsorship	−	−	0.48*	0.45
	−	−	(0.27)	(0.32)
Islamist Ideology	−	−	0.35	0.29
	−	−	(0.35)	(0.35)
Constant	−0.65	−1.02	0.56	0.25
	(0.68)	(0.66)	(0.41)	(0.43)
Observations	21959	27570	6768	8069

Note: *p < 0.10; **p < 0.05; ***p < 0.01 (two-tailed))

Logistic regression
Robust standard errors clustered on terrorist group in parentheses

Table A.21. Attack Claims by Taliban

	Excluding Taliban	Taliban Only
Civilian Target	−0.25***	−0.65**
	(0.03)	(0.11)
Constant	−1.77***	−0.32***
	(0.03)	(0.08)
Observations	28283	1525

Unit of analysis is terrorist attack
Logistic Regression
Robust standard errors, clustered on the terrorist group in parentheses

Table A.22. Learning to Deny Organizational Credit

	% Claimed		% Civilian Claimed	
	Model 1	Model 2	Model 3	Model 4
Age of Group	−0.008	−0.008	−0.008	−0.009
	(0.049)	(0.033)	(0.070)	(0.036)
Intercept	0.200	0.209	0.227	0.241
	(0.052)	(0.038)	(0.023)	(0.012)
Group FE	Yes	Yes	Yes	Yes
Max Age > 9	No	Yes	No	Yes
R2	0.562	0.555	0.619	0.629
Adjusted R2	0.467	0.494	0.522	0.566
N	320	260	256	207

Note: p-values (2-tailed) in parentheses.

Bibliography

Aaroncynic. "Why U.S. Governors Shouldn't—And Probably Won't—Be Able To Refuse Syrian Refugees." *Chicagoist,* November 17, 2015. http://chicagoist.com/2015/11/17/despite_declarations_governors_have.php.

Abi-Habib, Maria. "Syria's Civil War Produces a Clear Winner: Hezbollah." Wall Street Journal, April 3, 2017. https://www.wsj.com/articles/syrias-civil-war-produces-a-clear-winner-hezbollah-1491173790.

Abouzeid, Rania. "Interview with Official of Jabhat Al-Nusra, Syria's Islamist Militia Group." Time, December 25, 2012. http://world.time.com/2012/12/25/interview-with-a-newly-designated-syrias-jabhat-al-nusra/#ixzz2GWtcMfw9.

Abrahms, Max. "Al Qaeda's Scorecard: A Progress Report on Al Qaeda's Objectives." Studies in Conflict & Terrorism 29, no. 5 (August 1, 2006): 509–29.

Abrahms, Max. "Why Terrorism Does Not Work." *International Security* 31, no. 2 (October 11, 2006): 42–78.

Abrahms, Max. "Why Democracies Make Superior Counterterrorists." *Security Studies* 16, no. 2 (June 6, 2007): 223–53.

Abrahms, Max. "What Terrorists Really Want: Terrorist Motives and Counterterrorism Strategy." *International Security* 32, no. 4 (April 2008): 78–105.

Abrahms, Max. "Lumpers versus Splitters: A Pivotal Battle in the Field of Terrorism Studies." *Cato Unbound,* February 10, 2010. https://www.cato-unbound.org/2010/02/10/max-abrahms/lumpers-versus-splitters-pivotal-battle-field-terrorism-studies.

Abrahms, Max. "The Political Effectiveness of Terrorism Revisited." *Comparative Political Studies* 45, no. 3 (February 16, 2012): 366–93.

Abrahms, Max. "The Credibility Paradox: Violence as a Double-Edged Sword in International Politics." *International Studies Quarterly* 57, no. 4 (December 2013): 660–71.

Abrahms, Max. "Does #ISIS Care about Inducing Government Concessions?" Tweet. *Twitter,* September 2, 2014. https://twitter.com/MaxAbrahms/status/506864027663093760.

Abrahms, Max. "If #ISIS Wants More US Military Involvement in the Muslim World It Should Continue to Threaten the Homeland & Butcher American Journalists." Tweet. *Twitter,* August 19, 2014. https://twitter.com/MaxAbrahms/status/501848613686091777.

Abrahms, Max. "No, Civilian Targeting Isn't Politically Productive and #ISIS Leaders Will Increasingly Recognize That." Tweet. *Twitter,* September 1, 2014. https://twitter.com/MaxAbrahms/status/506547527085473792.

Abrahms, Max. "Note to #ISIS: Research Shows That Killing Hostages Only Lowers the Odds of Obtaining Ransoms and Other Concessions http://Tinyurl.Com/Mk6q25p Tweet. *Twitter,* August 26, 2014. https://twitter.com/MaxAbrahms/status/504307595218550784.

Abrahms, Max. "Research Shows Beheading Hostages like Foley, Sotloff, Haines, & Henning Is Counterproductive for the Perpetrators." Tweet. *Twitter*, October 1, 2014. https://twitter.com/MaxAbrahms/status/518134912235601921.

Abrahms, Max. "Sooner or Later the Entire World Is Going to Gang up and Crush #ISIS." Tweet. *Twitter*, August 4, 2014. https://twitter.com/MaxAbrahms/status/496413439439278080.

Abrahms, Max. "Terrorism Expert Max Abrahms on Islamic State," Interview. 2014. https://www.youtube.com/watch?v=-VlXIHfVqoo.

Abrahms, Max. "The Indiscriminate Violence of #ISIS Will Backfire, Research Shows." Tweet. *Twitter*, September 26, 2014. https://twitter.com/MaxAbrahms/status/515506985090162688.

Abrahms, Max. "The Main Reason to Oppose the #ISISmediaBlackout." Tweet. *Twitter*, September 2, 2014. https://twitter.com/MaxAbrahms/status/506968879156174848.

Abrahms, Max. "This Is Exactly Why I Opposed #ISISmediaBlackout: http://Bit.Ly/1qV4Ys4. The Grisly Violence Is Actually Counterproductive for the Group." Tweet. *Twitter*, September 1, 2014. https://twitter.com/MaxAbrahms/status/506526583851868162.

Abrahms, Max. "Will ISIS Brutality Backfire?" Tweet. *Twitter*, September 12, 2014. https://twitter.com/MaxAbrahms/status/515405592987709442.

Abrahms, Max. "Why the Islamic State Actually Stinks at Social Media." *OpenCanada*, April 20, 2015. https://www.opencanada.org/features/why-the-islamic-state-actually-stinks-at-social-media/.

Abrahms, Max. "Why People Keep Saying, 'That's What the Terrorists Want.'" *Harvard Business Review*, November 20, 2015.

Abrahms, Max, and Philip B.K. Potter. "Explaining Terrorism: Leadership Deficits and Militant Group Tactics." International Organization 69, no. 02 (2015): 311–42.

Abrahms, Max. "The T-word: When is an Attack Terrorism?" *Los Angeles Times*, November 8, 2017.

Abrahms, Max. "Correspondence: The Extremist's Disadvantage." *International Security* (Fall 2018).

Abrahms, Max, Nicholas Beauchamp, and Joseph Mroszczyk. "What Terrorist Leaders Want: A Content Analysis of Terrorist Propaganda Videos." *Studies in Conflict & Terrorism* 40, no. 11 (November 2, 2017): 899–916.

Abrahms, Max, and Justin Conrad. "The Strategic Logic of Credit Claiming: A New Theory for Anonymous Terrorist Attacks." *Security Studies* 26, no. 2 (April 3, 2017): 279–304.

Abrahms, Max, and Jochen Mierau. "Leadership Matters: The Effects of Targeted Killings on Militant Group Tactics." *Terrorism and Political Violence* 29, no. 5 (2017): 830–51.

Abrahms, Max, Matthew Ward, and Ryan Kennedy. "Explaining Civilian Attacks: Terrorist Networks, Principal-Agent Problems and Target Selection." *Perspectives on Terrorism* 12, no. 1 (2018).

Abū 'Amr, Ziyād. *Islamic Fundamentalism in the West Bank and Gaza: Muslim Brotherhood and Islamic Jihad*. Indiana Series in Arab and Islamic Studies. Bloomington: Indiana University Press, 1994.

Afghan, Wahdat. "Afghan Vice-Presidential Candidate Survives Ambush." *Reuters*, July 26, 2016. World News. https://www.reuters.com/article/us-afghanistan-attack/afghan-vice-presidential-candidate-survives-ambush-idUSTRE56P0R220090726.

Agnes, Melissa. "Timelessness of A Social Media Crisis Plan." *Melissa Agnes—Crisis Management Keynote Speaker*, March 22, 2012. http://melissaagnes.com/dominos-pizza-a-look-at-the-timelessness-of-a-social-media-crisis-plan/.

Agren, David. "Mexico after El Chapo: New Generation Fights for Control of the Cartel." *The Guardian*, May 5, 2017. World News. http://www.theguardian.com/world/2017/may/05/el-chapo-sinaloa-drug-cartel-mexico.

Ahmed, Azam, and Matthew Rosenberg. "Taliban Deny Responsibility for Attack on Red Cross." *The New York Times*, May 31, 2013. Asia Pacific. https://www.nytimes.com/2013/06/01/world/asia/taliban-deny-responsibility-for-attack-on-red-cross.html.

Al Baghdadi, Abu Bakr. "Abu Bakr Al Baghdadi Speech to Mosul (English Subtitles)." *YouTube.Com*, September 5, 2014. https://www.youtube.com/watch?v=qXNgNy8VX84.

"Al Qaeda Offers 'Condolences' for Innocent Victims." *CNN*, December 13, 2006. World. http://www.cnn.com/2009/WORLD/asiapcf/12/12/afghanistan.alqaeda/index.html.

Alfred, Charlotte. "What's Behind The Islamic State's Propaganda War." *Huffington Post*, October 11, 2015. https://www.huffingtonpost.com/entry/isis-propaganda-quilliam-foundation_us_56181d92e4b0dbb8000e9e45.

Alinsky, Saul David. *Rules for Radicals: A Practical Primer for Realistic Radicals*. 1st ed. New York: Random House, 1971.

Alkhshali, Hamdi, Phil Gast, and Barbara Starr. "US, Iraq Say ISIS Blew up Famous Mosul Mosque." *CNN*, June 23, 2017. World. http://www.cnn.com/2017/06/21/world/mosul-iraq-mosque-destroyed/index.html.

Allemann, Fritz René. "Terrorism: Definitional Aspects." *Terrorism* 3, no. 3–4 (January 1980): 185–90.

Allen, Lori. "Palestinians Debate 'Polite' Resistance to Occupation." *Middle East Report*, 32, no. 225 (Winter 2002): 38–43.

Almukhtar, Sarah. "How Boko Haram Courted and Joined the Islamic State." *The New York Times*, June 10, 2015. World. https://www.nytimes.com/interactive/2015/06/11/world/africa/boko-haram-isis-propaganda-video-nigeria.html.

"Al-Qaeda Urges More Attacks, Al-Zawahiri Tape." *Agence France Presse*, May 21, 2003. http://www.aljazeera.info/News%20archives/2003.html.

Al-Rahman, Atiyah Abd. "`Atiyah's Letter to Zarqawi (English Translation)." Combating Terrorism Center at West Point, December 12, 2005. https://ctc.usma.edu/posts/atiyahs-letter-to-zarqawi-english-translation-2.

Al-Shishani, Murad Batal. "The Amman Bombings: A Blow to the Jihadists?" *The Jamestown Foundation*, November 29, 2005. https://jamestown.org/program/the-amman-bombings-a-blow-to-the-jihadists/.

Al-Tamimi, Aymenn Jawad. "Archive of Islamic State Administrative Documents." *Aymenn Jawad Al-Tamimi*, January 27, 2015. http://www.aymennjawad.org/2015/01/archive-of-islamic-state-administrative-documents.

Altindal, Ahmet. "As Fighters Flee Isis We Should Be Worried about the Group They're about to Join." *The Independent*, August 24, 2017. http://www.independent.co.uk/voices/isis-hayy-at-tahrir-al-sham-future-of-syria-should-be-worried-a7909971.html?amp.

Al-Zawahiri, Ayman. "Zawahiri's Letter to Zarqawi (English Translation)." Combating Terrorism Center at West Point, July 9, 2005. https://ctc.usma.edu/posts/zawahiris-letter-to-zarqawi-english-translation-2.

Amarasingam, Amarnath. "Tahrir As-Sham Officially Denies They Were behind Damascus Bombing Today and Says Their Focus Is Only on Military Targets." Tweet. *Twitter*, March 15, 2017. https://twitter.com/AmarAmarasingam/status/842077495352561664.

Amir, Menachem. *Patterns in Forcible Rape*. Chicago: University of Chicago Press, 1971.

"An Assessment of the Nigerian Terrorist Group Boko Haram." *Bipartisan Policy Center*, May 15, 2014. https://bipartisanpolicy.org/blog/nigerian-terrorist-group-boko-haram/.

Anderson, Sulome. "What I Learned About Terrorism by Talking With My Father's Kidnapper." *New York Magazine—Daily Intelligencer*, October 26, 2016. http://nymag.com/daily/intelligencer/2016/10/what-i-learned-about-terror-from-my-fathers-kidnapper.html.

Andrews, Natalie, and Felicia Schwartz. "Islamic State Pushes Social-Media Battle With West." *Wall Street Journal*, August 22, 2014. World. http://www.wsj.com/articles/isis-pushes-social-media-battle-with-west-1408725614.

Animal Liberation Front. "The ALF Credo and Guidelines." *Animal Liberation Front*. Accessed October 5, 2017. http://www.animalliberationfront.com/ALFront/alf_credo.htm.

Ardolino, Bill and Bill Roggio, "New Details Emerge about Complex Attack On FOB Salerno." *Long War Journal*, June 10, 2012.

Arian, Alan. *Israeli Public Opinion on National Security, 2000*. Tel Aviv: Jaffee Center for Strategic Studies, Tel Aviv University, 2000. http://www.inss.org.il/upload/(FILE)1190278199.pdf.

Armstrong, Paul, and Hamdi Alkhshali. "Rebels in Syria Behead Boy in 'Mistake.'" *CNN*, July 21, 2016. Middle East. http://www.cnn.com/2016/07/20/middleeast/boy-beheaded-in-syria/index.html.

Arquilla, John, and David F. Ronfeldt, eds. *Networks and Netwars: The Future of Terror, Crime, and Militancy*. Santa Monica, CA: RAND, 2001.

Art, Robert J., and Patrick M. Cronin, eds. *The United States and Coercive Diplomacy*. Washington, D.C: United States Institute of Peace Press, 2003.

Asal, Victor H., R. Karl Rethemeyer, Ian Anderson, Allyson Stein, Jeffrey Rizzo, and Matthew Rozea. "The Softest of Targets: A Study on Terrorist Target Selection." *Journal of Applied Security Research* 4, no. 3 (July 17, 2009): 258–78.

Associated Press. "'Our Leaders Betrayed Us': Fear, Sectarianism behind Iraqi Army Collapse." *CBC News*, June 13, 2014. http://www.cbc.ca/news/world/iraqi-attacks-fear-sectarianism-behind-iraq-army-collapse-1.2674361.

Associated Press. "Nepal: Rebels Apologize For Bus Attack." *The New York Times*, June 8, 2005. World. https://www.nytimes.com/2005/06/08/world/world-business-briefing-asia-nepal-rebels-apologize-for-bus-attack.html.

Associated Press. "Nepalese Rebels Apologize for Deadly Bombing of Civilian Bus." *USA Today*, June 7, 2005. http://usatoday30.usatoday.com/news/world/2005-06-07-nepal-rebels_x.htm.

Associated Press. "UN: Boko Haram Use of Kids as 'Human Bombs' Soars in 2017." *AP News*, August 22, 2017. https://apnews.com/8ab141b808f7439a9a7a27e136c634ed.

Associated Press, and TOI Staff. "Syria's Al-Qaeda Affiliate Says It Regrets Killing of Druze." *The Times of Israel*, June 13, 2015. http://www.timesofisrael.com/syrias-al-qaeda-affiliate-says-it-regrets-killing-of-druze/.

Atran, Scott, and Nafees Hamid. "Paris: The War ISIS Wants." *The New York Review of Books*, November 16, 2015. http://www.nybooks.com/daily/2015/11/16/paris-attacks-isis-strategy-chaos/.

Aust, Stefan. *The Baader-Meinhof Group: The inside Story of a Phenomenon.* London: Bodley Head, 1987.

Aventajado, Roberto N., and Montelibano, Teodoro Y. *140 Days Of Terror: In The Clutches Of The Abu Sayyaf.* Pasig City, Philippines: Anvil Publishing, Inc., 2004.

Axelrod, Robert, ed. *Structure of Decision: The Cognitive Maps of Political Elites.* Princeton, NJ: Princeton University Pres, 1976.

Aznar, José María. "America and Europe After Bush." Freeman Spogli Institute, November 17, 2008. http://www.law.stanford.edu/calendar/details/2201/#related_information_and_recordings.

Barber, James David. *The Presidential Character: Predicting Performance in the White House.* 4th ed. Englewood Cliffs, N.J: Prentice Hall, 1992.

Barclay, Eliza. "With Lawsuit Over, Taco Bell's Mystery Meat Is A Mystery No Longer." *NPR*, April 19, 2011. Your Health. http://www.npr.org/sections/health-shots/2011/04/22/135539926/with-lawsuit-over-taco-bells-mystery-meat-is-a-mystery-no-longer.

Barghouti, Marwan. "Want Security? End the Occupation." *Washington Post*, January 16, 2002. https://www.washingtonpost.com/archive/opinions/2002/01/16/want-security-end-the-occupation/6d95b7aa-48bd-43e8-9698-e35331460ffb/.

Barker, Bernard. "Do Leaders Matter?" *Educational Review* 53, no. 1 (February 2001): 65–76.

Barker, Colin, Alan Johnson, and Michael Lavalette, eds. "Leadership Matters: An Introduction." In *Leadership and Social Movements.* Manchester, UK: Manchester University Press, 2001.

Barnard, Anne. "After Losses in Syria and Iraq, ISIS Moves the Goal Posts." *The New York Times*, October 18, 2016. Middle East. https://www.nytimes.com/2016/10/19/world/middleeast/islamic-state-syria-iraq.html.

Barnard, Anne, and Neil MacFarquhar. "Paris and Mali Attacks Expose Lethal Qaeda-ISIS Rivalry." *The New York Times*, November 20, 2015. Middle East. https://www.nytimes.com/2015/11/21/world/middleeast/paris-and-mali-attacks-expose-a-lethal-al-qaeda-isis-rivalry.html.

Barnea, Nahum. "We Want to Liberate You." *Yediot Ahronot.* September 2, 2001. https://www.freezerbox.com/archive/article.php?id=166.

Barnett, Donald. *The Revolution in Angola: Mpla, Life Histories and Documents.* Indianapolis, IN: Bobbs-Merrill Co, 1972.

Barnett, Donald, and Karari Njama. *Mau Mau from within: Autobiography and Analysis of Kenya's Peasant Revolt.* New York: Monthly Review Press, 1966.

Basil, Yousuf, and Catherine E. Shoichet. "Al Qaeda: We're Sorry about Yemen Hospital Attack." *CNN*, December 22, 2013. http://www.cnn.com/2013/12/22/world/meast/yemen-al-qaeda-apology/index.html.

Bassiouni, M. Cherif. *The Shariʿa and Islamic Criminal Justice in Time of War and Peace.* New York: Cambridge University Press, 2014.

Battersby, John D. "A.N.C. Acts to Halt Civilian Attacks." *The New York Times*, August 21, 1988. World. http://www.nytimes.com/1988/08/21/world/anc-acts-to-halt-civilian-attacks.html.

"Battle for Mosul: IS 'Blows up' Al-Nuri Mosque." *BBC News*, June 22, 2017. Middle East. http://www.bbc.com/news/world-middle-east-40361857.

Baumann, Michael. *Terror or Love? Bommi Baumann's Own Story of His Life as a West German Urban Guerrilla.* 1st ed. New York: Grove Press, 1979.

Baumeister, Roy F. *Evil: Inside Human Violence and Cruelty.* 1st Holt paperback ed. New York: Holt Comp, 1999.

Bazzi, Mohamad. "Commentary: Little Known Jihadist Inspired Latest Wave of 'Lone Wolf' Attacks." *Reuters*, July 28, 2016. Commentary. https://www.reuters.com/article/us-lone-wolf-attacks-commentary/commentary-jihadist-who-wrote-1600-page-manifesto-inspired-wave-of-lone-wolf-attacks-idUSKCN1071OS.

Beam, Louis. "Leaderless Resistance." *The Seditionist*, no. 12 (February 1992). http://www.louisbeam.com/leaderless.htm.

Becker, Michael. "Explaining Lone Wolf Target Selection in the United States." *Studies in Conflict & Terrorism* 37, no. 11 (November 2, 2014): 959–78.

Becker, Michael. "Why Violence Abates: Imposed and Elective Declines in Terrorist Attacks." *Terrorism and Political Violence* 29, no. 2 (March 4, 2017): 215–35.

Begin, Menachem. *The Revolt.* New York: Dell Publishing, 1978.

Benoit, William L. *Accounts, Excuses, and Apologies: A Theory of Image Restoration Strategies.* SUNY Series in Speech Communication. Albany: State University of New York Press, 1995.

Benoit, William L., and Shirley Drew. "Appropriateness and Effectiveness of Image Repair Strategies." *Communication Reports* 10, no. 2 (March 1997): 153–63.

Benotman, Norman, and Roisin Blake. "Jabhat al-Nusra: A Strategic Briefing." London: Quilliam Foundation, 2013.

Bergen, Peter. "What Empowers ISIS." *Herald Democrat*, November 6, 2015. Opinion. http://www.heralddemocrat.com/article/20151106/OPINION/311069974.

Bergen, Peter. "What Empowers ISIS." *CNN.* Opinions. Accessed October 5, 2017. http://www.cnn.com/2015/11/05/opinions/bergen-isis-on-a-roll/index.html.

Bergen, Peter. *The Osama Bin Laden I Know: An Oral History of Al-Qaeda's Leader.* 1st ed. New York: Free Press, 2006.

Berger, J. M., and Jonathon Morgan. "The ISIS Twitter Census: Defining and Describing the Population of ISIS Supporters on Twitter." Analysis Paper. The Brookings Project on U.S. Relations with the Islamic World, March 2015. https://www.brookings.edu/wp-content/uploads/2016/06/isis_twitter_census_berger_morgan.pdf.

Berinsky, Adam J., Gregory A. Huber, and Gabriel S. Lenz. "Evaluating Online Labor Markets for Experimental Research: Amazon.Com's Mechanical Turk." *Political Analysis* 20, no. 03 (2012): 351–68.

Berrebi, Claude. "The Economics of Terrorism and Counterterrorism: What Matters and Is Rational-Choice Theory Helpful?" In *Social Science for Counterterrorism: Putting the Pieces Together*, edited by Paul K. Davis and Kim Cragin, 151–208. Santa Monica, CA: RAND, 2009.

Berrebi, Claude, and Esteban F. Klor. "On Terrorism and Electoral Outcomes: Theory and Evidence from the Israeli-Palestinian Conflict." *Journal of Conflict Resolution* 50, no. 6 (December 2006): 899–925.

Berrebi, Claude, and Esteban F. Klor. "Are Voters Sensitive to Terrorism? Direct Evidence from the Israeli Electorate." *American Political Science Review* 102, no. 03 (August 1, 2008): 279–301.

Berrebi, Claude, and Esteban F. Klor. "On Terrorism and Electoral Outcomes. Theory and Evidence from the Israeli-Palestinian Conflict." *Journal of Conflict Resolution* 50, no. 6 (December 2006): 899–925.

Bershidsky, Leonid. "Islamic State Remains Dangerous in Defeat." *Bloomberg.com*, July 11, 2017. Opinion. https://www.bloomberg.com/view/articles/2017-07-11/islamic-state-remains-dangerous-in-defeat.

Best, Geoffrey. *War and Law since 1945*. Reprinted. Oxford, UK: Clarendon, 1997.

Bhasin, Kelly. "9 PR Fiascos That Were Handled Brilliantly By Management." *Business Insider*, May 26, 2011. http://www.businessinsider.com/pr-disasters-crisis-management-2011-5/#johnson-and-johnsons-cyanide-laced-tylenol-capsules-1982-1.

Bhojani, Fatima. "Despite Losing Terrain, Islamic State's Attacks Rose in 2016: Study." *Reuters*, August 21, 2017. World News. https://www.reuters.com/article/us-islamic-state-attacks/despite-losing-terrain-islamic-states-attacks-rose-in-2016-study-idUSKCN1B115Q.

Bin Laden, Osama. "Dear Muslim Brothers and Sisters," March 1, 2016. Office of the Director of National Intelligence. https://www.dni.gov/files/documents/ubl2016/english/Dear%20Muslim%20brothers%20and%20sisters.pdf.

Bin Laden, Osama. "Declaration of Jihad Against Americans Occupying the Land of the Two Holy Mosques." Azzam Publications, 1996. http://college.cengage.com/history/primary_sources/world/two_holy_mosques.htm.

Bin Laden, Osama. Interview with Osama bin Laden. Interview by Peter Bergen. Video, May 10, 1997. http://news.findlaw.com/cnn/docs/binladen/binladenintvw-cnn.pdf.

Bin Laden, Osama. "World Islamic Front Statement Urging Jihad Against Jews and Crusaders." Federation of American Scientists, February 23, 1998. https://fas.org/irp/world/para/docs/980223-fatwa.htm.

"Bin Laden Rails against Crusaders and UN." *BBC News*, November 3, 2001. http://news.bbc.co.uk/2/hi/world/monitoring/media_reports/1636782.stm.

"Bin Laden: 'Your Security Is in Your Own Hands.'" *CNN*, October 30, 2004. http://edition.cnn.com/2004/WORLD/meast/10/29/bin.laden.transcript/.

"Bin Laden's Warning: Full Text," October 7, 2001. http://news.bbc.co.uk/2/hi/south_asia/1585636.stm.

Bishara, Azmi. "The Quest for Strategy." *Journal of Palestine Studies* 32, no. 2 (January 2003): 41–9.

Bishop, Patrick, and Eamonn Mallie. *The Provisional IRA*. London: Corgi Books, 1992.

Blakeley, Georgina. "'It's Politics, Stupid!' The Spanish General Election of 2004." *Parliamentary Affairs* 59, no. 2 (February 10, 2006): 331–49.

Blandy, Charles W. "Military Aspects of the Two Russo–Chechen Conflicts in Recent Times." *Central Asian Survey* 22, no. 4 (December 2003): 421–32.

Blinken, Antony J. "The Islamic State Is Not Dead Yet." *The New York Times*, July 9, 2017. Opinion. https://www.nytimes.com/2017/07/09/opinion/islamic-state-mosul-iraq-strategy.html.

Blua, Antoine. "At Least 40 Killed In Kandahar Blast." *RadioFreeEurope/RadioLiberty*, June 10, 2010. News. https://www.rferl.org/a/Thirty_Nine_Killed_In_Kandahar_Blast/2066996.html.

Bob, Yonah. "Analysis: As Caliphate Falls, Isis Rewrites Its Playbook." *The Jerusalem Post*, November 17, 2016. Middle East. http://www.jpost.com/Middle-East/ISIS-Threat/Analysis-As-Caliphate-falls-ISIS-rewrites-its-playbook-472973.

Boot, Max. *Invisible Armies: An Epic History of Guerrilla Warfare from Ancient Times to the Present*. 1. ed. New York, NY: Norton, 2013.

Booth, Ken. "The Human Faces of Terror: Reflections in a Cracked Looking-Glass." *Critical Studies on Terrorism* 1, no. 1 (March 5, 2008): 65–79.

Bourgois, Philippe I. *In Search of Respect: Selling Crack in El Barrio*. 2nd ed. Structural Analysis in the Social Sciences 10. Cambridge: Cambridge University Press, 2003.

Bowker, Mike. "Russia and Chechnya: The Issue of Secession." *Nations and Nationalism* 10, no. 4 (October 1, 2004): 461–78.

Boyle, Darren. "ISIS Propaganda Machine Is Hit Hard by Drone Strikes." *Daily Mail Online*, June 11, 2017. http://www.dailymail.co.uk/news/article-4592898/ISIS-propaganda-machine-hit-hard-drone-strikes.html#ixzz4jjKQ5l6U.

Bradley, Ed. "Hezbollah: 'A-Team Of Terrorists,'" April 18, 2003. 60 Minutes. https://www.cbsnews.com/news/hezbollah-a-team-of-terrorists/.

Brandt, Patrick T., and Todd Sandler. "Hostage Taking: Understanding Terrorism Event Dynamics." *Journal of Policy Modeling* 31, no. 5 (September 2009): 758–78.

Brannen, Kate. "Hagel: ISIS Is More Dangerous Than Al Qaeda." *Foreign Policy*, August 21, 2014. https://foreignpolicy.com/2014/08/21/hagel-isis-is-more-dangerous-than-al-qaeda/.

Braungart, Richard G., and Margaret M. Braungart. "From Protest to Terrorism: The Case of SDS and the Weathermen." *International Social Movement Research* 4, no. 1 (1992): 45–78.

Breitenbücher, Danielle. "Syria, Code of Conduct of the Free Syrian Army: How Does Law Protect in War?" *International Committee of the Red Cross*. Accessed August 6, 2016. https://casebook.icrc.org/case-study/syria-code-conduct-free-syrian-army.

Brinson, Susan L., and William L. Benoit. "The Tarnished Star: Restoring Texaco's Damaged Public Image." *Management Communication Quarterly* 12, no. 4 (May 1999): 483–510.

Brown, Joseph. "The Bomber Who Calls Ahead: Pre-Attack Warnings as Helpful Threats." Presented at the 2017 Annual Conference of the International Studies Association. Baltimore, MD: 2017.

Brown, Joseph M. "Notes to the Underground: Credit Claiming and Organizing in the Earth Liberation Front." *Terrorism and Political Violence*, September 7, 2017, 1–20.

Browne, Ryan, and Barbara Starr. "US General Confirms Ending of Program to Arm Anti-Assad Rebels in Syria." *CNN*, July 21, 2017. Politics. http://www.cnn.com/2017/07/21/politics/us-stops-arming-anti-assad-rebels/index.html.

Brum, Pablo. "Revisiting Urban Guerrillas: Armed Propaganda and the Insurgency of Uruguay's MLN-Tupamaros, 1969–70." *Studies in Conflict & Terrorism* 37, no. 5 (May 4, 2014): 387–404.

Burke, Daniel. "Why ISIS Is Celebrating Trump's Immigration Ban." *CNN*, January 31, 2017. http://www.cnn.com/2017/01/31/us/islamerica-excerpt-grayzones/index.html.

Bush, George W. "President Bush: 'No Nation Can Be Neutral in This Conflict.'" Speech presented at the Warsaw Conference on Combatting Terrorism, New York, NY, November 6, 2001. https://georgewbush-whitehouse.archives.gov/news/releases/2001/11/20011106-2.html.

Bush, George W. "President Bush Speaks to United Nations." Speech presented at the United Nations General Assembly, New York, NY, November 10, 2001. https://georgewbush-whitehouse.archives.gov/news/releases/2001/11/20011110-3.html.

Bush, George W. "President Bush Speaks to United Nations." *The White House Archives,* November 10, 2001. https://georgewbush-whitehouse.archives.gov/news/releases/2001/11/20011110-3.html.

Bush, George W. "President's Address to Congress and the Nation on Terrorism." Speech presented at the Joint Session of Congress, Washington, DC, September 20, 2001. http://www.johnstonsarchive.net/terrorism/bush911c.html.

Bush, George W. "Presidential Speech to the California Business Association." Speech, Sacramento, CA, November 17, 2001.

Bush, George W. "Remarks to Employees of the Dixie Printing and Packaging Corporation." Speech, Glen Burnie, Maryland, October 24, 2001. http://www.presidency.ucsb.edu/ws/index.php?pid=64182.

Bush, George W. "President's Remarks at the United Nations General Assembly." Speech presented at the United Nations General Assembly, New York, NY, September 12, 2002. https://georgewbush-whitehouse.archives.gov/news/releases/2002/09/20020912-1.html.

Bush, George W. "Bush's Farewell Speech: 'This Is a Moment of Hope and Pride.'" Speech. *CNN,* February 16, 2009. http://www.cnn.com/2009/POLITICS/01/15/bush.speech.text/.

"Bush's War." *FRONTLINE.* PBS, March 24, 2008. http://www.pbs.org/wgbh/pages/frontline/bushswar/.

Byman, Daniel L., and Kenneth M. Pollack. "Let Us Now Praise Great Men: Bringing the Statesman Back In." *International Security* 25, no. 4 (April 2001): 107–46.

Byman, Daniel. "Taliban vs. Predator." *Foreign Affairs,* March 18, 2009. https://www.foreignaffairs.com/articles/south-asia/2009-03-18/taliban-vs-predator.

Byman, Daniel. *A High Price: The Triumphs and Failures of Israeli Counterterrorism.* Oxford, UK: Oxford University Press, 2011.

Byman, Daniel. "Why Drones Work: The Case for Washington's Weapon of Choice." *Brookings,* June 17, 2013. https://www.brookings.edu/articles/why-drones-work-the-case-for-washingtons-weapon-of-choice/.

Byman, Daniel. "Buddies or Burdens? Understanding the Al Qaeda Relationship with Its Affiliate Organizations." *Security Studies* 23, no. 3 (July 3, 2014): 431–70.

Byrd, Norman. "Child Beheading Video: US-Backed Syrian Rebel Group Video Execution Of Young Boy 'Spy.'" *Inquisitr,* July 22, 2016. https://www.inquisitr.com/3336176/child-beheading-video-us-backed-syrian-rebel-group-video-execution-of-young-boy-spy/.

Byrne, Paul. "Manchester Bomber Was Urged to 'Show No Mercy' by Online Islamic Extremists Ahead of Deadly Terror Attack." *Mirror,* August 14, 2017. News. http://www.mirror.co.uk/news/uk-news/manchester-bomber-salman-abedi-urged-10989063.

Callimachi, Rukmini. "Clues on Twitter Show Ties Between Texas Gunman and ISIS Network." *The New York Times,* May 11, 2015. U.S. https://www.nytimes.com/2015/05/12/us/twitter-clues-show-ties-between-isis-and-garland-texas-gunman.html.

Callimachi, Rukmini. "From ISIS' Standpoint, This Is a Success." Tweet. *Twitter,* September 15, 2017. https://twitter.com/rcallimachi/status/908770239915155457.

Callimachi, Rukmini. "How ISIS Built the Machinery of Terror Under Europe's Gaze." *The New York Times*, March 29, 2016. Europe. https://www.nytimes.com/2016/03/29/world/europe/isis-attacks-paris-brussels.html.

Cambanis, Thanassis. *A Privilege to Die*. London: Simon & Schuster, 2011.

Canter, David. *Criminal Shadows: Inside the Mind of the Serial Killer*. New York: Harper Collins, 1994.

Carter, Chelsea J. "ISIS Beheading U.S. Journalist James Foley, Posts Video." *CNN*, August 20, 2014. World. http://www.cnn.com/2014/08/19/world/meast/isis-james-foley/index.html.

Chalmers, David Mark. *Hooded Americanism: The History of the Ku Klux Klan*. 3rd ed. Durham: Duke University Press, 1981.

Chari, Raj. "The 2004 Spanish Election: Terrorism as a Catalyst for Change?" *West European Politics* 27, no. 5 (November 1, 2004): 954–63.

Chenoweth, Erica, Nicholas Miller, Elizabeth McClellan, Hillel Frisch, Paul Staniland, and Max Abrahms. "What Makes Terrorists Tick." *International Security* 33, no. 4 (2009): 180–202.

Chenoweth, Erica, and Maria J. Stephan. *Why Civil Resistance Works: The Strategic Logic of Nonviolent Conflict*. Paperback ed. Columbia Studies in Terrorism and Irregular Warfare. New York: Columbia University Press, 2011.

Chorley, Katharine. *Armies and the Art of Revolution*. Boston: Beacon Press, 1973.

Chotiner, Isaac. "The ISIS Correspondent." *Slate*, July 12, 2016. http://www.slate.com/articles/news_and_politics/interrogation/2016/07/rukmini_callimachi_the_new_york_times_isis_reporter_discusses_her_beat.html.

Chowanietz, Christophe. "Rallying around the Flag or Railing against the Government? Political Parties' Reactions to Terrorist Acts." *Party Politics* 17, no. 5 (September 2011): 673–98.

Chulov, Martin. "Isis Surrenders Iraqi Hideout of Leader Abu Bakr Al-Baghdadi." *The Guardian*, June 4, 2017. World News. http://www.theguardian.com/world/2017/jun/04/iraqi-forces-retake-key-town-of-baaj-from-isis.

Chulov, Martin. "Losing Ground, Fighters and Morale – Is It All over for Isis?" *The Guardian*, September 7, 2016. World News. http://www.theguardian.com/world/2016/sep/07/losing-ground-fighter-morale-is-it-all-over-for-isis-syria-turkey.

Cillizza, Chris. "These Trump Supporters Think Charlottesville Was a False Flag Operation." *CNN*, August 23, 2017. Politics. http://www.cnn.com/2017/08/23/politics/trump-focus-group/index.html.

"Civilian Casualties in Afghanistan up 14 per Cent Last Year, Says New UN Report." *UN News Service*, February 8, 2014. http://www.un.org/apps/news/story.asp?NewsID=47107#.WdgmjGhSxPY.

Clarke, Colin, and Daveed Gartenstein-Ross. "How Will Jihadist Strategy Evolve as the Islamic State Declines?" *War on the Rocks*, November 10, 2016. https://warontherocks.com/2016/11/how-will-jihadist-strategy-evolve-as-the-islamic-state-declines/.

Clark, Kate, and Borhan Osman. "First Wave of IS Attacks? Claim and Denial over the Jalalabad Bombs." *Afghanistan Analysts Network*, April 22, 2015. https://www.afghanistan-analysts.org/first-wave-of-is-attacks-claim-and-denial-over-the-jalalabad-bombs/.

Clarke, Liam, and Kathryn Johnston. *Martin McGuinness: From Guns to Government.* Edinburgh: Mainstream, 2001.

Clarke, Richard A. *Against All Enemies: Inside America's War on Terror.* 1st ed. New York: Free Press, 2004.

Clausewitz, Carl von. *On War.* Translated by James John Graham. New York: Routledge, 2017.

Clinton, William. "Clinton Statement in Full." Speech. *BBC News*, August 26, 1998. http://news.bbc.co.uk/2/hi/americas/155412.stm.

Coady, C. A. J. *Morality and Political Violence.* Cambridge, UK: Cambridge University Press, 2008.

Cocks, Tim. "Boko Haram Too Extreme for 'Al Qaeda in West Africa' BRAND." *Reuters*, May 28, 2014. World News. https://www.reuters.com/article/us-nigeria-bokoharam-analysis/boko-haram-too-extreme-for-al-qaeda-in-west-africa-bRAND-idUSKBN0E81D320140528.

Cohen, Dara Kay. *Rape during Civil War.* Ithaca: Cornell University Press, 2016.

Collins, Aukai. *My Jihad: The True Story of an American Mujahid's Amazing Journey from Usama Bin Laden's Training Camps to Counterterrorism with the FBI and CIA.* Guilford, Conn: Lyons Press, 2002.

Collins, Eamon, and Mick McGovern. *Killing Rage.* New York: Granta Books, 1998.

Collins, Randall. "Social Movements and the Focus of Emotional Attention." In *Passionate Politics: Emotions and Social Movements*, edited by Jeff Goodwin and James M. Jasper. Chicago: University of Chicago Press, 2001.

"Communist Rebels 'Sorry' for Circus Bombing in Davao City." *Inquirer*, September 7, 2012. http://newsinfo.inquirer.net/266026/communist-rebels-sorry-for-circus-bombing-in-davao-city.

Connolly, Kevin. "How the Omagh Case Unravelled." *BBC News*, December 20, 2007. Northern Ireland. http://news.bbc.co.uk/nol/ukfs_news/mobile/newsid_7150000/newsid_7154900/7154952.stm.

Contorno, Steve. "What Obama Said about Islamic State as a 'JV' Team." *Politifact*, September 7, 2014. http://www.politifact.com/truth-o-meter/statements/2014/sep/07/barack-obama/what-obama-said-about-islamic-state-jv-team/.

Conway, Kieran. *Southside Provisional: From Freedom Fighter to the Four Courts.* Dublin, Ireland: Orpen Press, 2014.

Cooke, Charles. "To Avoid Letting the Terrorists Win, Make Them Lose." *National Review*, November 16, 2015. http://www.nationalreview.com/article/427168/avoid-letting-terrorists-win-make-them-lose-charles-c-w-cooke.

Cordes, Bonnie, Bruce Hoffman, Brian Jenkins, Konrad Kellen, Sue Moran, and William Sater, eds. *Trends in International Terrorism, 1982 and 1983.* Santa Monica, CA: RAND, 1984.

Cordesman, Anthony H., and Sam Khazai. *Iraq in Crisis.* CSIS Reports. Center for Strategic & International Studies, 2014.

Corlett, J. Angelo. *Terrorism: A Philosophical Analysis.* Vol. 101. Philosophical Studies Series. Dordrecht: Springer Netherlands, 2003. http://public.eblib.com/choice/publicfullrecord.aspx?p=3103070.

Cottee, Simon. "No, the Travel Ban Isn't Being Used as ISIS Propaganda." *POLITICO*, March 26, 2017. http://www.politico.com/magazine/story/2017/03/travel-ban-isis-propaganda-214953.

Council on Foreign Relations. "The Taliban: A CFR InfoGuide Presentation." January 21, 2015. https://www.cfr.org/taliban.

Crary, David. "Experts Confront Multiple Explanations for Surge of Killings." *AP News*, July 28, 2016. Top News. https://apnews.com/ec0a2db52e9d43a299b920e605ec05b1/experts-confront-multiple-explanations-surge-killings.

Crawford, Jamie, and Laura Koran. "U.S. Officials: Foreigners Flock to Fight for ISIS." *CNN*, February 11, 2015. Politics. http://www.cnn.com/2015/02/10/politics/isis-foreign-fighters-combat/index.html.

Crenshaw, Martha. "The Causes of Terrorism." *Comparative Politics* 13, no. 4 (July 1981): 379.

Crenshaw, Martha, ed. *Terrorism, Legitimacy, and Power: The Consequences of Political Violence*. Middletown, CT: Wesleyan University Press, 1983.

Crenshaw, Martha. "Theories of Terrorism: Instrumental and Organizational Approaches." *Journal of Strategic Studies* 10, no. 4 (December 1987): 13–31.

Crenshaw, Martha. "Decisions to Use Terrorism: Psychological Constraints on Instrumental Reasoning." In *Social Movements and Violence: Participation in Underground Organizations*, edited by Donatella Della Porta, 29–42. Greenwich, CT: JAI Press, 1992.

Crenshaw, Martha. "The Logic of Terrorism." Terrorism in Perspective 24 (2007): 24–33.

Crenshaw, Martha. "The Debate over 'New' vs. 'Old' Terrorism." In *Values and Violence*, edited by Ibrahim A. Karawan, Wayne McCormack, and Stephen E. Reynolds, 4:117–36. Dordrecht: Springer Netherlands, 2008.

Cronin, Audrey Kurth. *How Terrorism Ends: Understanding the Decline and Demise of Terrorist Campaigns*. Princeton, NJ: Princeton Univ. Press, 2011.

Cronin, Audrey Kurth "ISIS Is Not a Terrorist Group." *Foreign Affairs*, October 5, 2016. https://www.foreignaffairs.com/articles/middle-east/isis-not-terrorist-group.

Crozier, Brian. *The Rebels: A Study of Post-War Insurrections*. London: Chatto and Windus, 1960.

Cunningham, Kathleen Gallagher. "Actor Fragmentation and Civil War Bargaining: How Internal Divisions Generate Civil Conflict." *American Journal of Political Science* 57, no. 3 (2013): 659–72.

Daraghmeh, Mohammed. "Militia Leader to Strengthen Army." *Plainview Daily Herald*. March 6, 2002. http://www.myplainview.com/news/article/Militia-Leader-to-Strengthen-Army-9012009.php.

Davis, James H., F. David Schoorman, and Lex Donaldson. "Toward a Stewardship Theory of Management." *Academy of Management Review* 22, no. 1 (1997): 20–47.

Dau, Luis Alfonso, Elizabeth M. Moore, and Max Abrahms. "Global Security Risks, Emerging Markets and Firm Responses: Assessing the Impact of Terrorism." In *Contemporary Issues in International Business*, pp. 79–97, Palgrave Macmillan, 2018.

Dearden, Lizzie. "Isis' 'Business Model' Failing as Group Haemorrhages Millions While Losing Territory across Syria and Iraq." *The Independent*, February 18, 2017. Middle East. http://www.independent.co.uk/news/world/middle-east/isis-islamic-state-daesh-funding-iraq-syria-territory-losses-oil-air-strikes-icsr-report-antiquities-a7586936.html.

Dearden, Lizzie. "Tens of Thousands of Refugees Are Returning to War-Torn Syria Because the World Is Failing Them." *The Independent*, July 1, 2017. http://www.independent.co.uk/news/world/middle-east/syria-civil-war-conflict-latest-refugees-returning-internally-displaced-thousands-half-million-un-a7817991.html.

Dearden, Lizzie. "Why Isis Fighters Are Convinced the Las Vegas Gunman Is One of Them." *The Independent.* October 6, 2017. Americas. http://www.independent.co.uk/news/world/americas/isis-vegas-shooting-stephen-paddock-repeat-claim-islamic-state-responsibility-police-gunman-motive-a7986161.html.

"Death Toll from France Truck Attack Rises to 85." *BNO News*, August 4, 2016. http://bnonews.com/news/index.php/news/id4998.

Decker, Scott H., and Barrik Van Winkle. *Life in the Gang: Family, Friends and Violence.* 1st ed. Cambridge Criminology Series. Cambridge, UK: Cambridge University Press, 1996.

Delaney, Paul. "Spain Fears Bombing May Herald an Increase in Terrorist Attacks." *The New York Times*, June 23, 1987. World. http://www.nytimes.com/1987/06/23/world/spain-fears-bombing-may-herald-an-increase-in-terrorist-attacks.html.

Della Porta, Donatella. "Left-Wing Terrorism in Italy." In *Terrorism in Context*, edited by Martha Crenshaw, 105–59. University Park, PA: Pennsylvania State University Press, 1995.

Della Porta, Donatella. *Social Movements, Political Violence, and the State: A Comparative Analysis of Italy and Germany.* 1st ed. Cambridge Studies in Comparative Politics. Cambridge, UK: Cambridge University Press, 2006.

Della Porta, Donatella, and Mario Diani, eds. *Social Movements: An Introduction.* 2nd ed. Boston: Wiley-Blackwell, 2006.

DeNardo, James. *Power in Numbers: The Political Strategy of Protest and Rebellion.* Princeton, N.J.: Princeton University Press, 1985.

De Roy van Zuijdewijn, Jeanine. "A Surge in 'Lone (or Loon) Wolf' Jihadist Attacks: The Future of Jihad? - Leiden Safety and Security Blog." *Leiden Safety and Security Blog*, January 5, 2015. http://leidensafetyandsecurityblog.nl/articles/a-surge-in-lone-or-loon-wolf-jihadist-attacks-the-future-of-jihad.

Dershowitz, Alan M. *Why Terrorism Works: Understanding the Threat, Responding to the Challenge.* New Haven: Yale University Press, 2002.

Dhawan, Gopi Nath. *The Political Philosophy of Mahatma Gandhi.* Bombay: The Popular Books Depot, 1946.

Diamond, Jeremy. "Trump Calls Manchester Attack Perpetrators 'Evil Losers.'" *CNN*, May 23, 2017. Politics. http://www.cnn.com/2017/05/23/politics/trump-manchester-remarks/index.html.

DiChristopher, Tom. "Islamic State Territory Is Shrinking, Taking Major Bite out of Its Financing." *CNBC*, March 31, 2017. Wars and Military Conflicts. https://www.cnbc.com/2017/03/31/islamic-state-territory-is-shrinking-taking-major-bite-out-of-funding.html.

Dlugy, Yana. "Putin Calls on Religious Leaders to Aid in Anti-Terror Fight." *Agence France-Presse*, December 29, 2004.

"Donald Trump: The President ISIS Wants?" MSNBC, May 24, 2016. http://www.msnbc.com/the-last-word/watch/donald-trump-the-president-isis-wants-692261443855.

Donnelly, Dónal. *Prisoner 1082: Escape from Crumlin Road Prison Europe's Alcatraz.* Cork: Collins Press, 2010.

Donovan, Tim. "PM: Election Will Go Ahead as Planned." *BBC News*, June 4, 2017. Election 2017. http://www.bbc.com/news/election-2017-40148918.

Downes, Alexander B. "Draining the Sea by Filling the Graves: Investigating the Effectiveness of Indiscriminate Violence as a Counterinsurgency Strategy." *Civil Wars 9*, no. 4 (December 2007): 420–44.

Drake, C. J. M. *Terrorists' Target Selection*. New York: Palgrave Macmillan, 1998.

Drakos, Konstantinos, and Andreas Gofas. "The Devil You Know but Are Afraid to Face: Underreporting Bias and Its Distorting Effects on the Study of Terrorism." *Journal of Conflict Resolution* 50, no. 5 (October 2006): 714–35.

Duka, Norman, Dennis Mercer, and Ginger Mercer. *From Shantytown to Forest, The Story of Norman Duka*. Vol. 1. LSM Information Center, 1974.

DuPee, Matthew C., Thomas H. Johnson, and Matthew P. Dearing. "Understanding Afghan Culture: Analyzing the Taliban Code of Conduct: Reinventing the Layeha." Culture & Conflict Studies Occasional Paper Series. Monterey, CA: Department of National Security Affairs, Naval Postgraduate School, August 6, 2009. https://info. publicintelligence.net/Layeha.pdf.

Dzutsev, Valery. "Leader of the Caucasus Emirate Vows to Stop Attacks against Russian Civilians." *North Caucasus Analysis* 13, no. 3 (February 3, 2012). https://jamestown.org/ program/leader-of-the-caucasus-emirate-vows-to-stop-attacks-against-russian-civilians/.

Earthmann, Art. "The Date '9/11' (2001)...Chosen To Avenge Muslim Defeat At Battle Of Vienna On '9/11' (1583)!" *Thom Hartmann*, June 23, 2014. https://www.thomhartmann. com/users/art-earthmann/blog/2014/06/date-911-2001-chosen-avenge-muslim-defeat-battle-vienna-911-1583.

Editor of Encyclopaedia Britannica. "Hezbollah | Lebanese Militant Group & Political Party." *Encyclopedia Britannica*, April 7, 2017. https://www.britannica.com/topic/ Hezbollah.

Eilstrup-Sangiovanni, Mette, and Calvert Jones. "Assessing the Dangers of Illicit Networks: Why Al-Qaida May Be Less Threatening Than Many Think." *International Security* 33, no. 2 (2008): 7–44.

Eisenhardt, Kathleen M. "Agency Theory: An Assessment and Review." *Academy of Management Review* 14, no. 1 (January 1, 1989): 57–74.

Eldar, Shlomi. *Eyeless in Gaza*. Tel Aviv: Yediot Aharonot, 2005.

El-Ghobashy, Tamer, Mustafa Salim, and Louisa Loveluck. "Islamic State's Caliphate Has Been Toppled in Iraq and Syria. Why Isn't Anyone Celebrating?" *Washington Post*, December 5, 2017. https://www.washingtonpost.com/world/middle_east/islamic-statescaliphate-has-been-toppled-in-iraq-and-syria-why-isnt-anyone-celebrating/2017/ 12/04/67737794-c97a-11e7-b506-8a10ed11ecf5_story.html?tid=ss_tw&utm_term=. b970f47e44a0

Ellis, Ralph. "KKK Rally in Charlottesville Outnumbered by Counterprotesters." *CNN*, July 10, 2017. http://www.cnn.com/2017/07/08/us/kkk-rally-charlottesville-statues/ index.html.

Ellis, Stephen. *The Mask of Anarchy: The Destruction of Liberia and the Religious Dimension of an African Civil War*. 2nd ed. New York: New York University Press, 2006.

Enders, Walter, and Todd Sandler. "What Do We Know about the Substitution Effect in Transnational Terrorism?" In *Research on Terrorism: Trends, Achievements & Failures*, edited by Andrew Silke, 119–37. Cass Series on Political Violence. London: Routledge, 2004.

Enders, Walter, and Todd Sandler. *The Political Economy of Terrorism*. 2nd ed. Cambridge, UK: Cambridge University Press, 2012.

Engel, Pamela. "ISIS Has Mastered a Crucial Recruiting Tactic No Terrorist Group Has Ever Conquered." *Business Insider.* May 9, 2015. http://www.businessinsider.com/isis-is-revolutionizing-international-terrorism-2015-5.

Engel, Pamela. "'It's Similar to North Korea': Inside ISIS's Sophisticated Strategy to Brainwash People in the 'Caliphate.'" *Business Insider,* November 28, 2015. http://www.businessinsider.com/isis-propaganda-strategy-2015-11.

Engel, Pamela. "ISIS' Caliphate Is Shrinking, and the Terror Group Is about to Lose One of Its Biggest Cities." *Business Insider,* October 17, 2016. http://www.businessinsider.com/how-much-territory-has-isis-lost-2016-10.

English, Richard. *Does Terrorism Work?* 1st ed. Oxford: Oxford University Press, 2016.

Ertugrul Mavioglu. "Civilians in Turkey Off Target List, PKK Boss Says." *Hürriyet Daily-News.* October 28, 2010.

"Explosion Kills Syrian Rebel Leader." *BBC News,* September 10, 2014. Middle East. http://www.bbc.com/news/world-middle-east-29135922.

Fanon, Frantz, and Richard Philcox. *The Wretched of the Earth: Frantz Fanon ; Translated from the French by Richard Philcox; Introductions by Jean-Paul Sartre and Homi K. Bhabha.* New York: Grove Press, 2004.

"Fatalities in the First Intifada." *B'Tselem—The Israeli Information Center for Human Rights in the Occupied Territories.* Accessed August 3, 2005. http://www.btselem.org/statistics/first_intifada_tables.

Fellman, Michael. *The Guerilla Conflict in Missouri During the American Civil War.* New York: Oxford University Press, 1989.

Felter, Joseph, and Jarret Brachman. "An Assessment of 516 Combatant Status Review Tribunal Unclassified Summaries." CTC Report. West Point, NY: United States Military Academy Combating Terrorism Center, July 15, 2007.

Ferrell, O. C., and Steven J. Skinner. "Ethical Behavior and Bureaucratic Structure in Marketing Research Organizations." *Journal of Marketing Research* 25, no. 1 (February 1988): 103.

Fielding, David, and Madeline Penny. "What Causes Changes in Opinion about the Israeli-Palestinian Peace Process?" Economics Discussion Papers. University of Otago, March 2006. http://citeseerx.ist.psu.edu/viewdoc/download?doi=10.1.1.570.6164&rep=rep1&type=pdf. Findley, Michael G., and Joseph K. Young. "Terrorism, Democracy, and Credible Commitments1: Terrorism, Democracy, and Credible Commitments." *International Studies Quarterly* 55, no. 2 (June 2011): 357–78.

Findley, Michael G., and Joseph K. Young. "Terrorism, Democracy, and Credible Commitments1: Terrorism, Democracy, and Credible Commitments." *International Studies Quarterly* 55, no. 2 (June 2011): 357–78.

Findley, Michael G., and Joseph K. Young. "More Combatant Groups, More Terror?: Empirical Tests of an Outbidding Logic." *Terrorism and Political Violence* 24, no. 5 (November 1, 2012): 706–21.

Fisher, Max. "Al-Qaeda Faction in Syria Hands out Teletubbies and Spiderman Dolls." *Washington Post.* August 13, 2013. WorldViews. https://www.washingtonpost.com/news/worldviews/wp/2013/08/13/al-qaeda-faction-in-syria-hands-out-teletubbies-and-spiderman-dolls/.

237

Fisher, Max. "9 Questions about ISIS You Were Too Embarrassed to Ask - Vox." *Vox Media*, November 23, 2015. https://www.vox.com/2015/11/23/9779188/isis-syria-iraq-9-questions.

Fischer, Martin. "As Strongholds Fall ISIS Is Rushing to Funnel Money out of Syria and Iraq." *Business Insider*, June 7, 2017. http://www.businessinsider.com/as-strongholds-fall-isis-rushing-to-funnel-money-out-of-syria-and-iraq-2017-6.

Fishman, Brian. *The Master Plan: ISIS, Al Qaeda, and the Jihadi Strategy for Final Victory*. New Haven: Yale University Press, 2016.

Fisk, Robert. "As My Grocer Said: Thank You Mr. Clinton for the Fine Words." *The Independent*, August 22, 1998.

Fleeman, Michael. "FBI Chemist Testifies Explosives Residue Found on McVeigh's Clothing." *Indianapolis Star*, May 20, 1997.

Fletcher, Holly. *Al-Aqsa Martyrs Brigade*. Washington, DC: Council on Foreign Relations, 2005. https://www.cfr.org/backgrounder/al-aqsa-martyrs-brigade.

Foreign Broadcast Information Service. *Compilation of Usama Bin Ladin Statements: 1994–January 2004*. Washington, DC: U.S. Department of State, 2004. https://fas.org/irp/world/para/ubl-fbis.pdf.

Foroohar, Kambiz. "Islamic State's Finances Cut by 30% From Last Year, Report Finds." *Bloomberg.Com*, April 18, 2016. https://www.bloomberg.com/news/articles/2016-04-18/islamic-state-s-finances-cut-by-30-from-last-year-report-finds.

Fortna, Virginia Page. "Do Terrorists Win? Rebels' Use of Terrorism and Civil War Outcomes." *International Organization* 69, no. 03 (2015): 519–56.

Freeman, Colin. "Iraq Crisis: Baghdad's Shia Militia in Defiant 50,000-Strong Rally as Isis Make Further Gains." *The Telegraph*. June 21, 2014. World. http://www.telegraph.co.uk/news/worldnews/middleeast/iraq/10916926/Iraq-crisis-Baghdads-Shia-militia-in-defiant-50000-strong-rally-as-Isis-make-further-gains.html.

Freeman, John, Glenn R. Carroll, and Michael T. Hannan. "The Liability of Newness: Age Dependence in Organizational Death Rates." *American Sociological Review* 48, no. 5 (October 1983): 692–710.

Friend, Celeste. "Social Contract Theory." *Internet Encyclopedia of Philosophy*. Accessed October 5, 2017. http://www.iep.utm.edu/soc-cont/.

"Full Text: Bin Laden's 'Letter to America.'" *The Guardian*, November 24, 2002. Observer Worldview. http://www.theguardian.com/world/2002/nov/24/theobserver.

Fullerton, Andrew, Nicole Renfer, and Erin Weiler. "TEXACO Case Study: Racial Discrimination Case," March 5, 2013. https://erincweiler.files.wordpress.com/2013/03/texaco-case-study.pdf.

Gacemi, Baya. *I, Nadia, Wife of a Terrorist*. Translated by Paul Cote. Lincoln, NE: University of Nebraska Press, 2006.

Gaddis, John Lewis. *Strategies of Containment: A Critical Appraisal of American National Security Policy during the Cold War*. Rev. and expanded ed. Oxford, UK: Oxford University Press, 2005.

Galbraith, Jay R. "Organization Design." In *Handbook of Organization Development*, edited by Thomas G. Cummings, 325–52. Thousand Oaks, CA: SAGE Publications, 2008.

Gambill, Gary C. "The Balance of Terror: War by Other Means in the Contemporary Middle East." *Journal of Palestine Studies* 28, no. 1 (October 1998): 51–66.

Gamson, William A. *The Strategy of Social Protest*. The Dorsey Series in Sociology. Homewood, IL: Dorsey Press, 1975.

Gandhi, Mahatma. Autobiography: *The Story of my Experiments with Truth*. Chelmsford, MA: Courier Corporation, 1948.

Garcia, O. Krumme. "Security over Freedom – Or: How Europe Defeats Itself against Terrorism." *Vocal Europe*, November 25, 2015. http://www.vocaleurope.eu/security-over-freedom-or-how-europe-defeats-itself-against-terrorism/.

Gardner, David. "The Hideous Dialectic of Isis Savagery." *Financial Times*, February 18, 2015. Isis. https://www.ft.com/content/de2135f6-b772-11e4-981d-00144feab7de.

Garrow, David J. *Bearing the Cross: Martin Luther King, Jr., and the Southern Christian Leadership Conference*. 1st ed. New York: W. Morrow, 1986.

Gebeily, Maya. "How ISIL Is Gaming the World's Journalists." *Public Radio International*, June 26, 2014. Conflict & Justice. https://www.pri.org/stories/2014-06-26/how-isil-gaming-worlds-journalists.

George, Alexander L., and William E. Simons, eds. *The Limits of Coercive Diplomacy*. 2nd ed. Boulder, CO: Westview Press, 1994.

Gerges, Fawaz. "The Strategic Logic of the Islamic State." *The National Interest—The Buzz*, August 14, 2016. http://nationalinterest.org/blog/the-buzz/the-strategic-logic-the-islamic-state-17351.

Gerges, Fawaz. *The Rise and Fall of Al-Qaeda*. Oxford, UK: Oxford University Press, 2014.

Gerges, Fawaz A. *ISIS: A History*. Princeton, NJ: Princeton University Press, 2016. https://doi.org/10.1515/9781400880362.

Getmansky, Anna, and Thomas Zeitzoff. "Terrorism and Voting: The Effect of Rocket Threat on Voting in Israeli Elections." *American Political Science Review* 108, no. 03 (August 2014): 588–604.

Getmansky, Anna, and Tolga Sinmazdemir. "Settling on Violence: Expansion of Israeli Outposts in the West Bank in Response to Terrorism." *Studies in Conflict & Terrorism* 41, no. 3 (2018): 241–259.

Ghatwai, Milind. "2013 Naxal Attack: CPI (Maoist) Leadership Regrets Killing Congress Leaders." *The Indian Express*, May 31, 2015. http://indianexpress.com/article/india/india-others/2013-naxal-attack-cpi-maoist-leadership-regrets-killing-congress-leaders/.

Giáp, Vo Nguyên. *People's War, People's Army*. Honolulu: University Press of the Pacific, 2001.

Gill, Paul, John Horgan, and Paige Deckert. "Bombing Alone: Tracing the Motivations and Antecedent Behaviors of Lone-Actor Terrorists." *Journal of Forensic Sciences* 59, no. 2 (March 2014): 425–35.

Gleis, Joshua L., and Benedetta Berti. *Hezbollah and Hamas: A Comparative Study*. Baltimore: Johns Hopkins University Press, 2012.

Goffman, Erving. "Remedial Interchanges." In *Relations in Public: Microstudies of the Public Order*, 95–187. New York, NY: Harper & Row, 1972.

Goldsmith, Jett. "How the Islamic State's Massive PR Campaign Secured Its Rise." *Bellingcat*, February 11, 2015. https://www.bellingcat.com/news/mena/2015/02/11/how-the-islamic-states-pr-campaign-secured-its-rise/.

Gonzales, Marti H., Debra J. Manning, and Julie A. Haugen. "Explaining Our Sins: Factors Influencing Offender Accounts and Anticipated Victim Responses." *Journal of Personality and Social Psychology* 62, no. 6 (1992): 958–71. doi:10.1037/0022-3514.62.6.958.

Goodwin, Jeff. "A Theory of Categorical Terrorism." *Social Forces* 84, no. 4 (2006): 2027–46.

Gopal, Anand. "The Taliban in Kandahar." In *Talibanistan: Negotiating the Borders between Terror, Politics and Religion*, edited by Peter L. Bergen and Katherine Tiedemann. Oxford, UK: Oxford University Press, 2013.

Gordon, Philip H. "Madrid Bombings and U.S. Policy" Testimony to U.S. Senate Foreign Relations Committee, March 31, 2004. http://www.senate.gov/?foreign/testimony/2004/GordonTestimony040331.pdf.

Gorenberg, Gershom. "Trump as a Strategic Asset of the Islamic State." *The American Prospect*, June 15, 2016. http://prospect.org/article/trump-strategic-asset-islamic-state.

Gould, Eric. D., and Esteban. F. Klor. "Does Terrorism Work?" *The Quarterly Journal of Economics* 125, no. 4 (November 1, 2010): 1459–1510.

Gould, Erica R. "Delegating IMF Conditionality: Understanding Variations in Control and Conformity." In *Delegation and Agency in International Organizations*, edited by Darren G. Hawkins, David A. Lake, Daniel L. Nielson, and Michael J. Tierney, 281–311. Cambridge: Cambridge University Press, 2006. doi:10.1017/CBO9780511491368.011.

Grathwohl, Larry, and Frank Reagan. *Bringing down America: An FBI Informer with the Weathermen*. 1st ed. New York, NY: Arlington House, 1976.

Gray, Alfred M. "Forward." In *Peacekeepers at War: Beirut 1983—the Marine Commander Tells His Story*, 1st ed. Washington, D.C: Potomac Books, 2009.

Greenberg, Joel. "Mideast Turmoil: Palestinian; Suicide Planner Expresses Joy Over His Missions." *The New York Times*, May 9, 2002. World. https://www.nytimes.com/2002/05/09/world/mideast-turmoil-palestinian-suicide-planner-expresses-joy-over-his-missions.html.

Greene, Thomas H. *Comparative Revolutionary Movements: Search for Theory and Justice*. 2nd ed. Prentice-Hall Contemporary Comparative Politics Series. Englewood Cliffs, NJ: Prentice-Hall, 1990.

Gregg, Richard B. *Power of Non-Violence*. New York: Schocken Books, 1966. http://www.myilibrary.com?id=891302.

Grint, Keith. *The Arts of Leadership*. Oxford, UK: Oxford University Press, 2000.

Guevara, Che. *Guerrilla Warfare*. Lanham, MD: Rowman & Littlefield Publishers, 2002.

Gunter, Michael M. *The Kurds and the Future of Turkey*. New York: Palgrave Macmillan, 1997.

Gurr, Ted R. "Some Characteristics of Political Terrorism in the 1960s." In *The Politics of Terrorism*, 33:31–53. Public Administration and Public Policy. Boca Raton, FL: CRC Press, 1979. https://www.ncjrs.gov/App/Publications/abstract.aspx?ID=56210.

Gurr, Ted R. "Empirical Research on Political Terrorism: The State of the Art and How It Might Be Improved." In *Current Perspectives on International Terrorism*, edited by Robert O. Slater and Michael Stohl, 115–54. London: Palgrave Macmillan, 1988. doi:10.1007/978-1-349-18989-2_5.

Gurr, Ted Robert. Political Rebellion: Causes, Outcomes and Alternatives. London: Routledge, 2015.

Guttman, Louis, "Public Assessment of the Activities and Violence of the Intifada." *Israel Institute of Applied Social Research*, December 1990.

Ha, Tu Thanh. "FLQ Terrorist Convicted of Killing Cabinet Minister Pierre Laporte Has Died." *The Globe and Mail*, January 15, 2015. Canada. https://www.theglobeandmail.

com/news/national/flq-terrorist-convicted-of-killing-cabinet-minister-pierre-laporte-has-died/article22466196/.

Hadar, Leon T. "Pakistan in America's War against Terrorism: Strategic Ally or Unreliable Client?" *Policy Analysis*. Washington, DC: CATO Institute, May 8, 2002. https://www.cato.org/publications/policy-analysis/pakistan-americas-war-against-terrorism-strategic-ally-or-unreliable-client.

Hafez, Mohammed M. "Armed Islamist Movements and Political Violence in Algeria." *Middle East Journal* 54, no. 4 (2000): 572–91.

Hafez, Mohammed M. *Suicide Bombers in Iraq: The Strategy and Ideology of Martyrdom*. Washington, DC: U.S. Institute of Peace Press, 2007.

Hall, Todd, and Keren Yarhi-Milo. "The Personal Touch: Leaders' Impressions, Costly Signaling, and Assessments of Sincerity." *International Studies Quarterly* 56, no. 3 (September 2012): 560–73.

Hall, Valerie, Hugh Mackay, and Colin Morgan. *Headteachers at Work*. Milton Keynes, UK: Open University Press, 1986.

Hamas. "The Covenant of the Islamic Resistance Movement." The Avalon Project, August 18, 1988. http://avalon.law.yale.edu/20th_century/hamas.asp.

Hamid, Shadi. "Is There a Method to ISIS's Madness?" *Brookings—Markaz*, November 24, 2015. https://www.brookings.edu/blog/markaz/2015/11/24/is-there-a-method-to-isiss-madness/.

Hamid, Shadi. *Islamic Exceptionalism: How the Struggle over Islam Is Reshaping the World*. First edition. New York: St. Martin's Press, 2016.

Hammami, Rema, and Jamil Hilal. "An Uprising at a Crossroads." Middle East Research and Information Project. Middle East Research and Information Project, Summer 2001. http://www.merip.org/mer/mer219/uprising-crossroads.

Hamzeh, Ahmad Nizar. *In the Path of Hizbullah*. 1st ed. Modern Intellectual and Political History of the Middle East. Syracuse, NY: Syracuse University Press, 2004.

Hanchett, Ian. "Alan Colmes: Calling ISIS 'Evil' Is 'What the Terrorists Want.'" *Breitbart*, August 25, 2014. http://www.breitbart.com/video/2014/08/25/alan-colmes-calling-isis-evil-what-the-terrorists-want/.

Harris, Gardiner, and Michael D. Shear. "Obama Says of Terrorist Threat: 'We Will Overcome It.'" *The New York Times*, December 6, 2015. Politics. https://www.nytimes.com/2015/12/07/us/politics/president-obama-terrorism-threat-speech-oval-office.html.

Harrison, Kim. "When Saying Sorry Is the Right Thing to Do." *Cutting Edge PR*. Accessed October 5, 2017. http://www.cuttingedgepr.com/articles/saying-sorry-is-right-to-communicate.asp.

Hart, Alan. *Arafat, a Political Biography*. Bloomington: Indiana University Press, 1989.

Hart, Robert A. "Democracy and the Successful Use of Economic Sanctions." *Political Research Quarterly* 53, no. 2 (June 2000): 267–84.

Hassan, Hassan. "Isis, the Jihadists Who Turned the Tables." *The Guardian*, August 9, 2014. The Observer. http://www.theguardian.com/world/2014/aug/10/isis-syria-iraq-barack-obama-airstrikes.

Hassan, Hassan. "The ISIS March Continues: From Ramadi on to Baghdad?" *Foreign Policy*, May 19, 2015. Argument. http://foreignpolicy.com/2015/05/19/ramadi-is-the-canary-isis-islamic-state-iraq/.

Hassan, Hassan. "Isis Has Reached New Depths of Depravity." *The Guardian*, February 8, 2015. The Observer. http://www.theguardian.com/world/2015/feb/08/isis-islamic-state-ideology-sharia-syria-iraq-jordan-pilot.

Hassan, Hassan. "Hassan Hassan on How to Defeat the Extremists." *The National*, July 3, 2017. https://www.pressreader.com/uae/the-national-news/20170703/282368334674447.

Hassan, Hassan. "What ISIL Really Thinks about the Future." *The National*, July 2, 2017. Opinion. https://www.thenational.ae/opinion/what-isil-really-thinks-about-the-future-1.91394.

Havard, Kate, and Jonathan Schanzer. "By Hosting Hamas, Qatar Is Whitewashing Terror." *Newsweek*, May 11, 2017. Opinion. http://www.newsweek.com/qatar-hosting-hamas-whitewashing-terror-606750.

Hawkins, Darren G., David A. Lake, Daniel L. Nielson, and Michael J. Tierney, eds. *Delegation and Agency in International Organizations*. Political Economy of Institutions and Decisions. Cambridge, UK: Cambridge University Press, 2006.

Heißner, Stefan, Peter R. Neumann, John Holland-McCowan, and Rajan Basra. "Caliphate in Decline: An Estimate of Islamic State's Financial Fortunes." London: The International Cenhhhhatre for the Study of Radicalisation and Political Violence, February 17, 2017. http://icsr.info/2017/02/icsrey-report-caliphate-decline-estimate-islamic-states-financial-fortunes/.

Hendawi, Hamza, and Abdul-Zahra Qassim. "IS Top Command Dominated by Ex-Officers in Saddam's Army." *Houston Chronicle*, August 9, 2015. http://www.houstonchronicle.com/news/nation-world/world/article/IS-top-command-dominated-by-ex-officers-in-6433282.php.

Hendrix, Cullen, and Joseph K. Young. "Weapon of the Weak?: Assessing the Effects of State Capacity on Terrorism," 1–32. Presented at Midwest Political Science Association Annual Meeting, Chicago, IL, 2012. https://www.google.com/url?sa=t&rct=j&q=&esrc=s&source=web&cd=3&cad=rja&uact=8&ved=0ahUKEwiwqMT_vuHWAhVh7YMKHTdqDWIQFggtMAI&url=http%3A%2F%2Ffs2.american.edu%2Fjyoung%2Fwww%2Fdocuments%2Fcap_terror_hy12.docx&usg=AOvVaw3uXj7GE3RssMYt95fiRC8j.

Heneghan, Tom. "ISIS's Intriguing Silence about Donald Trump's Approach to Muslims." *Religion News Service*, March 17, 2017. http://religionnews.com/2017/03/17/isis-intriguing-silence-about-donald-trumps-approach-to-muslims/.

Hensley, Nicole. "Charlottesville Suspect Held Hate Group Shield before Attack." *NY Daily News*, August 13, 2017. http://www.nydailynews.com/news/national/charlottesville-suspect-held-hate-group-shield-attack-article-1.3407245.

Herd, Graeme P. "Information Warfare and the Second Chechen Campaign." In *War & Peace in Post-Soviet Eastern Europe*, edited by Sally Cummings. Camberley, Surrey: Conflict Studies Research Centre, Royal Military Academy, Sandhurst, 2000.

Hewitt, Christopher. *Consequences of Political Violence*. Brookfield, VT: Dartmouth, 1993.

Hoffman, Aaron M. "Voice and Silence: Why Groups Take Credit for Acts of Terror." *Journal of Peace Research* 47, no. 5 (September 28, 2010): 615–26.

Hoffman, Bruce. "The Confluence of International and Domestic Trends in Terrorism." *Terrorism and Political Violence* 9, no. 2 (June 1997): 1–15.

Hoffman, Bruce. *Inside Terrorism*. New York: Columbia University Press, 1998.

Hoffman, Bruce. *Inside Terrorism*. Rev. and expanded ed. New York: Columbia University Press, 2016.

Hoffman, Bruce. "The Awful Truth Is That Terrorism Works." *The Daily Beast*, March 14, 2015. World. http://www.thedailybeast.com/articles/2015/03/14/the-awful-truth-is-that-terrorism-works.

Hoffman, Bruce. *Anonymous Soldiers: The Struggle for Israel, 1917–1947*. Reprint edition. New York: Vintage, 2016.

Hoffmann, Stanley. *Gulliver's Troubles: Or, the Setting of American Foreign Policy*. New York: McGraw-Hill, 1968.

Holland, Heidi. *The Struggle: A History of the African National Congress*. New ed. New York: Braziller, 1990.

Holland, Joshua. "Why Have a Record Number of Westerners Joined the Islamic State?" *BillMoyers.Com*, October 10, 2014. http://billmoyers.com/2014/10/10/record-number-westerners-joined-islamic-state-great-threat/.

Holsti, Ole R. "The Belief System and National Images: A Case Study." *The Journal of Conflict Resolution* 6, no. 3 (1962): 244–52.

Honig, Or, and Ariel Reichard. "The Usefulness of Examining Terrorists' Rhetoric for Understanding the Nature of Different Terror Groups." *Terrorism and Political Violence*, February 14, 2017, 1–20. doi:10.1080/09546553.2017.1283308.

"Hope for Chechnya." *The Economist*, June 1, 1996. https://www.highbeam.com/doc/1G1-18347488.html.

Horgan, John. *Walking Away from Terrorism: Accounts of Disengagement from Radical and Extremist Movements*. 1st ed. Political Violence. New York: Routledge, 2009.

Horowitz, Michael C., and Allan C. Stam. "How Prior Military Experience Influences the Future Militarized Behavior of Leaders." *International Organization* 68, no. 3 (2014): 527–59.

Horowitz, Michael C., Allan C. Stam, and Cali M. Ellis. *Why Leaders Fight*. Cambridge, UK: Cambridge University Press, 2015.

Horowitz, Michael C., and Philip B. K. Potter. "Allying to Kill: Terrorist Intergroup Cooperation and the Consequences for Lethality." *Journal of Conflict Resolution* 58, no. 2 (March 2014): 199–225.

Houston, G. F. *The National Liberation Struggle in South Africa: A Case Study of the United Democratic Front, 1893–1987*. Aldershot: Ashgate, 1999. http://repository.hsrc.ac.za/handle/20.500.11910/8508.

Huang, Yi-Hui. "Crisis Situations, Communication Strategies, and Media Coverage: A Multicase Study Revisiting the Communicative Response Model." *Communication Research* 33, no. 3 (June 2006): 180–205.

Hubbard, Ben, and Eric Schmitt. "Military Skill and Terrorist Technique Fuel Success of ISIS." *The New York Times*, August 27, 2014. Middle East. https://www.nytimes.com/2014/08/28/world/middleeast/army-know-how-seen-as-factor-in-isis-successes.html.

Hubbard, Ben, and Eric Schmitt. "ISIS, Despite Heavy Losses, Still Inspires Global Attacks." *The New York Times*, July 8, 2017. Middle East. https://www.nytimes.com/2017/07/08/world/middleeast/isis-syria-iraq.html.

Hudson, John, and Yochi Dreazen. "U.S. Won't Ship Iraq The Weapons It Needs to Fight Al Qaeda." *Foreign Policy*, January 7, 2014. https://foreignpolicy.com/2014/01/07/u-s-wont-ship-iraq-the-weapons-it-needs-to-fight-al-qaeda/.

Humphreys, Macartan, and Jeremy M. Weinstein. "Handling and Manhandling Civilians in Civil War." *American Political Science Review* 100, no. 03 (August 2006): 429–47.

"Hunting Bin Laden." *FRONTLINE*, May 1998. http://www.pbs.org/wgbh/pages/frontline/shows/binladen/who/interview.html.

Hutchinson, Ruth Gillette, Arthur R. Hutchinson, and Mabel Newcomer. "A Study in Business Mortality: Length of Life of Business Enterprises in Poughkeepsie, New York, 1843-1936." *The American Economic Review* 28, no. 3 (1938): 497–514.

Ibrahim, Saad Eddin. "Anatomy of Egypt's Militant Islamic Groups: Methodological Note and Preliminary Findings." *International Journal of Middle East Studies* 12, no. 4 (1980): 423–53.

Ibrahim, Waleed. "Al Qaeda's Two Top Iraq Leaders Killed in Raid." *Reuters*, April 19, 2010. World News. https://www.reuters.com/article/us-iraq-violence-alqaeda/iraq-says-local-al-qaeda-leader-has-been-killed-idUSTRE63I3CL20100419.

Ignatius, David. "The Manual That Chillingly Foreshadows the Islamic State." *Washington Post*, September 25, 2014. Opinion. https://www.washingtonpost.com/opinions/david-ignatius-the-mein-kampf-of-jihad/2014/09/25/4adbfc1a-44e8-11e4-9a15-137aa0153527_story.html.

Ignatius, David. "How the World Can Prepare for the 'Day After' ISIS." *Asharq Al-Awsat*, May 27, 2017. https://english.aawsat.com/david-ignatius/opinion/world-can-prepare-day-isis.

"In Recruitment Efforts, ISIS Seeks To Evoke Deep Sympathies From Muslims." *All Things Considered*. NPR, November 16, 2014. http://www.npr.org/2014/11/16/364545340/in-recruitment-efforts-isis-seeks-to-evoke-deep-sympathies-from-muslims.

International Crisis Group. "Who Governs the West Bank? Palestinian Administration under Israeli Occupation." Middle East Report. Amman: International Crisis Group, September 28, 2004. https://www.crisisgroup.org/middle-east-north-africa/eastern-mediterranean/israelpalestine/who-governs-west-bank-palestinian-administration-under-israeli-occupation.

IPS Correspondents. "PAKISTAN: Taliban Backs Off From Attacking Civilians." *Inter Press Service News Agency*, July 23, 2011. http://www.ipsnews.net/2011/07/pakistan-taliban-backs-off-from-attacking-civilians/.

"IRA Says Harrods Bombing Was a Mistake." *United Press International*, December 30, 1983. UPI Archives. https://www.upi.com/Archives/1983/12/30/IRA-says-Harrods-bombing-was-a-mistake/2270441608400/.

"'Iraq's Rambo' out of Favour with His Own Militia after Brutally Violent Videos." *France 24*, June 15, 2017. The Observers. http://observers.france24.com/en/20170615-iraq-rambo-out-favour-own-militia-after-brutally-violent-videos.

"ISIL Video Shows Christian Egyptians Beheaded in Libya." *Al Jazeera*, February 16, 2015. http://www.aljazeera.com/news/middleeast/2015/02/isil-video-execution-egyptian-christian-hostages-libya-150215193050277.html.

"Islamic State Group 'Lost Quarter of Territory in 2016.'" *BBC News*, January 19, 2017. Middle East. http://www.bbc.com/news/world-middle-east-38641509.

"Islamic State's Global Affiliates Interactive World Map." *IntelCenter*, December 5, 2015. https://intelcenter.com/maps/is-affiliates-map.html#gs.0kuWV2c.

"Islamic State's Media Violence May Hurt the Group," *VOA News.* August 3, 2015. https://learningenglish.voanews.com/a/islamic-states-media-violence-may-hurt-the-group/2888435.html.

"Islamist Sheikh Abu Osama Al-'Iraqi Denounces Al-Qaeda in Iraq for Atrocities against Sunnis." Special Dispatch. The Middle East Media Research Institute, November 1, 2006. https://www.memri.org/reports/islamist-sheikh-abu-osama-al-iraqi-denounces-al-qaeda-iraq-atrocities-against-sunnis.

Israel Democracy Institute. "Peace Index." *Tami Steinmetz Center for Peace Research, Tel Aviv University.* Accessed October 9, 2017. https://en.idi.org.il/centers/1159/1520.

Jackson, Brian A. "Groups, Networks, or Movements: A Command-and-Control-Driven Approach to Classifying Terrorist Organizations and Its Application to Al Qaeda." *Studies in Conflict & Terrorism* 29, no. 3 (May 2006): 241–62.

Jenkins, Brian, Janera Johnson, and David Ronfeldt. "Numbered Lives. Some Statistical Observations from 77 International Hostage Episodes,." Santa Monica, CA: RAND, July 1977. http://www.dtic.mil/docs/citations/ADA166288.

Jenkins, Brian M. "Statements about Terrorism." *The Annals of the American Academy of Political and Social Science* 463, no. 1 (September 1982): 11–23.

Jervis, Robert. *The Logic of Images in International Relations.* New York: Columbia University Press, 1989.

Jervis, Robert. "Do Leaders Matter and How Would We Know?" *Security Studies* 22, no. 2 (April 2013): 153–79.

Jo, Hyeran. *Compliant Rebels: Rebel Groups and International Law in World Politics.* Problems of International Politics. Cambridge, UK: Cambridge University Press, 2017.

Johns, Sheridan. "Obstacles to Guerrilla Warfare—A South African Case Study." *The Journal of Modern African Studies* 11, no. 2 (1973): 267–303.

Johnsen, Gregory D. "Securing Yemen's Cooperation in the Second Phase of the War on Al-Qa'ida." *CTC Sentinel*, December 15, 2007. https://ctc.usma.edu/posts/securing-yemen%e2%80%99s-cooperation-in-the-second-phase-of-the-war-on-al-qaida.

Johnson, Henry. "Mapped: The Islamic State Is Losing Its Territory—and Fast." *Foreign Policy*, March 16, 2016. http://foreignpolicy.com/2016/03/16/mapped-the-islamic-state-is-losing-its-territory-and-fast/.

Johnson, Thomas H., and Matthew C. DuPee. "Analysing the New Taliban Code of Conduct (Layeha): An Assessment of Changing Perspectives and Strategies of the Afghan Taliban." *Central Asian Survey* 31, no. 1 (February 16, 2012): 77–91.

Johnson, Tim. "Nice Massacre 'Arguably the Most Lethal Lone-Wolf Attack Ever.'" *McClatchy DC Bureau*, July 15, 2016. National. http://www.mcclatchydc.com/news/nation-world/national/article89903527.html.

Johnston, Patrick B. "Does Decapitation Work? Assessing the Effectiveness of Leadership Targeting in Counterinsurgency Campaigns." *International Security* 36, no. 4 (April 2012): 47–79.

Johnston, Patrick B., and Anoop K. Sarbahi. "The Impact of US Drone Strikes on Terrorism in Pakistan." *International Studies Quarterly* 60, no. 2 (June 2016): 203–19.

Jones, Benjamin F., and Benjamin A. Olken. "Do Leaders Matter? National Leadership and Growth Since World War II." *The Quarterly Journal of Economics* 120, no. 3 (August 1, 2005): 835–64.

Jones, David Martin. "Mad and Bad." *The Spectator*, August 13, 2016. Features. https://www.spectator.co.uk/2016/08/mad-and-bad/.

Jones, Edward E., and Daniel McGillis. "Correspondent Inferences and the Attribution Cube: A Comparative Reappraisal." In *New Directions in Attribution Research*, edited by John H. Harvey, William John Ickes, and Robert F. Kidd, 1:389–420. Hillsdale, NJ: L. Erlbaum Associates, 1976.

Jones, Edward E., and Keith E. Davis. "From Acts To Dispositions The Attribution Process In Person Perception." In *Advances in Experimental Social Psychology*, 2:219–66. New York: Academic Press, 1965. doi:10.1016/S0065-2601(08)60107-0.

Jones, Seth G., and Martin C. Libicki. *How Terrorist Groups End: Lessons for Countering Al Qa'ida*. Santa Monica, CA: RAND, 2008.

Joosse, Paul. "Leaderless Resistance and Ideological Inclusion: The Case of the Earth Liberation Front." *Terrorism and Political Violence* 19, no. 3 (July 4, 2007): 351–68.

Jordan, Jenna. "When Heads Roll: Assessing the Effectiveness of Leadership Decapitation." *Security Studies* 18, no. 4 (December 2, 2009): 719–55.

Joseph, Paul. *"Soft" Counterinsurgency: Human Terrain Teams and US*. New York: Palgrave Macmillan, 2017.

Kaag, John, and Sarah E. Kreps. *Drone Warfare*. 1st ed. War and Conflict in the Modern World. Cambridge, UK: Polity Press, 2014.

Kahn, Si. *Organizing: A Guide for Grassroots Leaders, Revised Edition*. Rev Sub ed. Silver Spring, MD: NASW Press, 1992.

Kahneman, D., and A. Tversky. "On the Reality of Cognitive Illusions." *Psychological Review* 103, no. 3 (July 1996): 582–91.

Kalyvas, Stathis N. *The Logic of Violence in Civil War*. 1st ed. Cambridge, UK: Cambridge University Press, 2006.

Kaplan, David E., and Andrew Marshall. *The Cult at the End of the World: The Incredible Story of Aum*. London: Hutchinson, 1996.

Kathrada, Ahmed. *Memoirs*. Reprint edition. Cape Town: Zebra Press, 2004.

Kaufman, Edy. "Israeli Perceptions of the Palestinians' 'Limited Violence' in the *Intifada*." *Terrorism and Political Violence* 3, no. 4 (December 1991): 1–38.

Kaufmann, Chaim. "Threat Inflation and the Failure of the Marketplace of Ideas: The Selling of the Iraq War." *International Security* 29, no. 1 (2004): 5–48.

Kavanaugh, Shane Dixon. "ISIS' Brutality In Videos Is 'Terrorist Clickbait.'" *Vocativ*, June 24, 2015. http://www.vocativ.com/world/isis-2/isis-escalating-brutality-in-videos-is-terrorist-clickbait/.

Kelly, Laura. "Robert Gates: Expect More Attacks in the West as ISIS Is Defeated in the Middle East." *The Washington Times*, May 23, 2017.//www.washingtontimes.com/news/2017/may/23/robert-gates-expect-more-attacks-west-isis-defeate/.

Kelly, Michael J., and Thomas H. Mitchell. "Transnational Terrorism and the Western Elite Press." *Political Communication* 1, no. 3 (January 1981): 269–96.

Khalilov, Roman. "The Russian-Chechen Conflict." *Central Asian Survey* 21, no. 4 (January 2002): 411–15.

Khan, Muqtedar. "Prospects for Muslim Democracy: The Role of U.S. Policy." *Middle East Policy* 10, no. 3 (September 1, 2003): 79–89.

Kilberg, Joshua. "A Basic Model Explaining Terrorist Group Organizational Structure." *Studies in Conflict & Terrorism* 35, no. 11 (November 2012): 810–30.

Klein, Malcolm W. *The American Street Gang: Its Nature, Prevalence, and Control.* Studies in Crime and Public Policy. New York: Oxford University Press, 1997.

Kleponis, Greg. "Throwing the Book at the Taliban." *Small Wars Journal*, September 1, 2010. http://smallwarsjournal.com/jrnl/art/throwing-the-book-at-the-taliban.

Klimova, S., and Public Opinion Foundation. "Negotiating with Maskhadov." *Фонд Общественное Мнение*, November 21, 2002.

Klimova, S., and Public Opinion Foundation. "Attitude to Chechens: Pity and Fear." *Фонд Общественное Мнение*, January 30, 2003.

Knight, Peter. "Outrageous Conspiracy Theories: Popular and Official Responses to 9/11 in Germany and the United States". *New German Critique*, no. 35 (2008): 165–93.

Koerner, Brendan. "Why ISIS Is Winning the Social Media War—And How to Fight Back." *WIRED*, April 2016. https://www.wired.com/2016/03/isis-winning-social-media-war-heres-beat/.

Kossoy, Edward. *Living with Guerrilla: Guerrilla as a Legal Problem and a Political Fact.* Geneva: Librairie Droz, 1976.

Kovaleski, Serge F., and Susan Schmidt. "Bombing's Repercussions Rattle Militias." *Washington Post*, May 6, 1995.

Krause, Peter. "The Political Effectiveness of Non-State Violence: A Two-Level Framework to Transform a Deceptive Debate." *Security Studies* 22, no. 2 (April 2013): 259–94.

Kun, Bela. "Discipline and Centralised Leadership." *The Communist Review* 3, no. 9 & 10 (January 1923). https://www.marxists.org/history/international/comintern/sections/britain/periodicals/communist_review/1923/09-10/dis_and_leader.htm.

Kydd, Andrew H., and Barbara F. Walter. "The Strategies of Terrorism." *International Security* 31, no. 1 (2006): 49–80.

Lago, Ignacio, and Jose Ramon Montero. "The 2004 Election in Spain: Terrorism, Accountability, and Voting." *Taiwan Journal of Democracy* 2, no. 1 (July 2006): 13–36.

Lahoud, Nelly. *Jihadis' Path to Self-Destruction.* 1st edition. Oxford: Oxford University Press, 2010.

Lake, David A. "Rational Extremism: Understanding Terrorism in the Twenty-First Century." *Dialogue IO* 1, no. 1 (Spring 2002): 15–28.

Lakey, George. *The Sociological Mechanisms of Nonviolent Action.* Philadelphia: University of Pennsylvania, 1962.

Lanktree, Graham. "Alt-Right 'America First' Rallies Move Online After Boston Free Speech Protest." *Newsweek*, August 22, 2017. http://www.newsweek.com/alt-right-america-first-rallies-move-online-after-boston-free-speech-protest-653372.

Lanoue, David J. "The 'Teflon Factor': Ronald Reagan & Comparative Presidential Popularity." *Polity* 21, no. 3 (Spring 1989): 481–501. doi:10.2307/3234744.

Lapan, Harvey E., and Todd Sandler. "Terrorism and Signalling." *European Journal of Political Economy* 9, no. 3 (August 1, 1993): 383–97.

Laqueur, Walter. "The Futility of Terrorism." *Harper's Magazine*, March 1976.

Laqueur, Walter. "Interpretations of Terrorism: Fact, Fiction and Political Science." *Journal of Contemporary History* 12, no. 1 (1977): 1–42.

Laqueur, Walter. *The Terrorism Reader—A Historical Anthology*. Philadelphia: Temple University Press, 1978.

Laqueur, Walter. *A History of Terrorism*. Piscataway, NJ: Transaction Publishers, 2016.

Laqueur, Walter, and Yonah Alexander, eds. *The Terrorism Reader: A Historical Anthology*. Philadelphia: Temple University Press, 1978.

Lasswell, Harold D., and Abraham Kaplan. *Power and Society: A Framework for Political Inquiry*. 1st ed. New Brunswick, N.J: Routledge, 2013.

Lee, Betty Kaman. "Audience-Oriented Approach to Crisis Communication: A Study of Hong Kong Consumers' Evaluation of an Organizational Crisis." *Communication Research* 31, no. 5 (October 1, 2004): 600–618.

Leon, Dan. "Israeli Public Opinion Polls on the Peace Process." *Palestine-Israel Journal* 2, no. 1 (1995). http://www.pij.org/details.php?id=676.

Lerner, Davide. "It's Not Islam That Drives Young Europeans to Jihad, France's Top Terrorism Expert Explains." *Haaretz*, August 20, 2017. Europe. https://www.haaretz.com/world-news/europe/1.791954.

Lesser, Ian O., Bruce Hoffman, John Arquilla, David Ronfeldt, Michele Zanini, and Brian Michael Jenkins. *Countering the New Terrorism*. Santa Monica, CA: RAND, 1999.

Levitt, Matthew. *Hezbollah: The Global Footprint of Lebanon's Party of God*. Hezbollah. Washington, DC: Georgetown University Press, 2013.

Levy, Gideon. "Death Isn't a Big Deal Anymore." *Ha'aretz*. November 8, 2001. http://www.bintjbeil.com/E/occupation/levy/011112.html.

Lewin, Hugh. *Stones Against the Mirror: Friendship in the Time of the South African Struggle*. 1st ed. Cape Town: Random House Struik, 2011.

Lewis, Aidan. "Islamic State Shifts to Libya's Desert Valleys after Sirte Defeat." *Reuters*, February 10, 2017. http://uk.reuters.com/article/uk-libya-security-islamicstate/islamic-state-shifts-to-libyas-desert-valleys-after-sirte-defeat-idUKKBN15P1H1.

Lewis, Danny. "Five Times the United States Officially Apologized." *Smithsonian*, May 27, 2016. Smart News. http://www.smithsonianmag.com/smart-news/five-times-united-states-officially-apologized-180959254/.

Lichbach, Mark. *The Rebel's Dilemma*. Ann Arbor: University of Michigan Press, 1998. https://www.press.umich.edu/13967/rebels_dilemma.

Lichfield, John. "Atrocity Is Sign Jihadists Are Losing Battle." *The Jewish Chronicle*, June 9, 2017. News. https://www.thejc.com/news/news-features/atrocity-is-sign-jihadists-are-losing-battle-1.439798.

Lind, Jennifer. *Sorry States: Apologies in International Politics*. Cornell Studies in Security Affairs. Ithaca, NY: Cornell University Press, 2008.

Lindo, Jason M., and María Padilla-Romo. "Kingpin Approaches to Fighting Crime and Community Violence: Evidence from Mexico's Drug War." Working Paper. National Bureau of Economic Research, May 2015. doi:10.3386/w21171.

Linschoten, Alex S. and Felix Kuehn. *Separating the Taliban from Al-Qaeda: The Core of Success in Afghanistan*. New York: Center for International Cooperation, New York University, 2011.

Lipsky, Michael. *Protest in City Politics: Rent Strikes, Housing, and the Power of the Poor*. Rand McNally, 1969.

Lister, Charles R. *The Syrian Jihad: Al-Qaeda, the Islamic State and the Evolution of an Insurgency.* 1st ed. Oxford, UK: Oxford University Press, 2016.

Lister, Tim, Ray Sanchez, Mark Bixler, Sean O'Key, Michael Hogenmiller, and Mohammed Tawfeeq. "ISIS: 143 Attacks in 29 Countries Have Killed 2,043." *CNN*, July 25, 2016. World. http://www.cnn.com/2015/12/17/world/mapping-isis-attacks-around-the-world/index.html.

Little, John. "Why Terrorism Fails: A Discussion with Max Abrahms." *Blogs of War*, March 16, 2015. http://blogsofwar.com/why-terrorism-fails-a-discussion-with-max-abrahms/.

Litvak, Meir. "The Islamization of the Palestinian-Israeli conflict: the case of Hamas." *Middle Eastern Studies* 34, no. 1 (1998): 148–163.

Livanios, Dimitris. "'Conquering the Souls': Nationalism and Greek Guerrilla Warfare in Ottoman Macedonia, 1904–1908." *Byzantine and Modern Greek Studies* 23, no. 1 (January 1, 1999): 195–221.

Lochhead, Carolyn. "Bush Insists on U.N. Resolve." *SFGate*. November 11, 2001. http://www.sfgate.com/politics/article/Bush-insists-on-U-N-resolve-U-S-pledges-boost-2857774.php.

Lochhead, Carolyn. "U.S. Pledges Boost in Aid to Pakistan." *San Francisco Chronicle*, November 11, 2001. http://www.sfgate.com/politics/article/Bush-insists-on-U-N-resolve-U-S-pledges-boost-2857774.php.

Lovell, David W., and Igor Primoratz, eds. *Protecting Civilians During Violent Conflict: Theoretical and Practical Issues for the 21st Century.* Military and Defence Ethics. Farnham, UK: Ashgate Publishing, 2012.

Lovley, Erika. "FBI Releases Murtha Files." *POLITICO*, May 25, 2010. http://www.politico.com/news/stories/0510/37757.html.

Lula, Karolina. "Why Defining Terrorism Matters." *The Monkey Cage*, May 28, 2013. http://themonkeycage.org/2013/05/why-defining-terrorism-matters/.

Macaraig, Ayee. "Philippine Casino Attacker a Gambling Addict, Not Terrorist: Police." *Yahoo News*, June 4, 2017. https://www.yahoo.com/news/philippine-casino-attacker-gambling-addict-not-terrorist-police-060904361.html.

MacCoun, Robert J., Elizabeth Kier, and Aaron Belkin. "Does Social Cohesion Determine Motivation in Combat? An Old Question with an Old Answer." *Armed Forces & Society* 32, no. 4 (July 1, 2006): 646–54.

MacDiarmid, Campbell. "From the Rise of the Caliphate to the Fall of Mosul, One Family's Journey." *Foreign Policy*, July 26, 2017. http://foreignpolicy.com/2017/07/26/from-the-rise-of-the-caliphate-to-the-fall-of-mosul-one-familys-journey-baghdadi-al-nuri-mosque-isis/.

Machlis, Avi. "Israel Hopes Fence Will Stop Suicide Bombers." *Financial Times*, June 16, 2002. www.globalpolicy.org/security/issues/israel-palestine/2002/0616wall.htm4

Mackintosh, Eliza. "Trump Ban Is Boon for ISIS Recruitment, Experts Say." *CNN*, January 31, 2017. Politics. http://www.cnn.com/2017/01/30/politics/trump-ban-boosts-isis-recruitment/index.html.

MacRae, Gordon. "In the Absence of Fathers: A Story of Elephants and Men." *These Stone Walls*, June 20, 2012. http://thesestonewalls.com/gordon-macrae/in-the-absence-of-fathers-a-story-of-elephants-and-men/.

Magnowski, Daniel. "Taliban Says Stop Civilian Deaths, but Actions Speak Louder." *Reuters*, November 8, 2011. World News. https://www.reuters.com/article/us-afghanistan-taliban-civilians/taliban-says-stop-civilian-deaths-but-actions-speak-louder- idUSTRE7A71ZK20111108.

Maher, Shiraz. *Salafi-Jihadism: The History of an Idea*. Oxford, UK: Oxford University Press, 2016.

Maher, Shiraz. "Shiraz Maher on Isis: The Management of Savagery." *New Statesman Magazine*, July 12, 2016. https://www.newstatesman.com/culture/2016/07/shiraz-maher-isis-management-savagery.

Mahoney, Charles W. "Splinters and Schisms: Rebel Group Fragmentation and the Durability of Insurgencies." *Terrorism and Political Violence* (2017): 1–20.

"Major Jihadi Cleric and Author of Al-Qaeda's Shari'a Guide to Jihad." Special Dispatch. The Middle East Media Research Institute, April 10, 2008. https://www.memri.org/.

Malet, David. *Foreign Fighters: Transnational Identity in Civil Conflicts*. 1st ed. Oxford, UK: Oxford University Press, 2013.

Mandela, Nelson. "I Am Prepared to Die: Nelson Mandela's Statement from the Dock at the Opening of the Defense Case in the Rivonia Trial." Speech, Pretoria, April 20, 1964. http://www.historyplace.com/speeches/mandela.htm.

Mandela, Nelson. *Long Walk to Freedom: The Autobiography of Nelson Mandela*. New York: Back Bay Books, 2013.

Mar, Roman. "Spain Won't Bow to Calls to Pull Troops." *Associated Press*, December 1, 2003.

Marco, Meg. "Burger King Employee Takes Bath In Sink, Feels Wrath Of Health Department." *Consumerist*, August 12, 2008. https://consumerist.com/2008/08/12/burger-king-employee-takes-bath-in-sink-feels-wrath-of-health-department/.

Marcus, Aliza. *Blood and Belief: The PKK and the Kurdish Fight for Independence*. 1. publ. in paperback. New York: New York University Press, 2009.

Marcus, Itamar, and Barbara Crook. "Arafat Blames Israel for Tel Aviv Bombing." *Palestinian Media Watch*, July 14, 2004. http://www.imra.org.il/story.php3?id=21474.

Margalit, Avishai. "The Suicide Bombers." *The New York Review of Books*, January 16, 2003. http://www.nybooks.com/articles/2003/01/16/the-suicide-bombers/.

Marighella, Carlos. *Minimanual of the Urban Guerrilla*. Seattle: Praetorian Press, 2011.

Marinov, Nikolay. "Do Economic Sanctions Destabilize Country Leaders?" *American Journal of Political Science* 49, no. 3 (2005): 564–76.

"Martin Place Siege: Joint Commonwealth—New South Wales Review." Canberra, AUS: Commonwealth of Australia, February 22, 2015. https://www.pmc.gov.au/sites/default/files/publications/170215_Martin_Place_Siege_Review_1.pdf.

Mashtaq, Anooshe. "A Weaker ISIS In The Middle East Means A Stronger Threat Abroad." *Huffington Post*, September 6, 2017. http://www.huffingtonpost.com.au/anooshe-mushtaq/a-weaker-isis-in-the-middle-east-means-a-stronger-threat-abroad_a_22133515/.

Matesan, Ioana Emy, and Ronit Berger. "Blunders and Blame: How Armed Non-State Actors React to Their Mistakes." *Studies in Conflict & Terrorism* 40, no. 5 (May 4, 2017): 376–98.

McAdam, Doug. *Freedom Summer*. Oxford, UK: Oxford University Press, 1990.

McCants, William. "How ISIL Out-Terrorized Bin Laden." *POLITICO*, August 19, 2015. http://politi.co/1Nm78RB.

McCants, William. "How the Islamic State's Favorite Strategy Book Explains Recent Terrorist Attacks." *War on the Rocks*, November 24, 2015. https://warontherocks.com/2015/11/how-the-islamic-states-favorite-strategy-book-explains-recent-terrorist- attacks/.

McCants, William. *The ISIS Apocalypse: The History, Strategy, and Doomsday Vision of the Islamic State*. 1st ed. New York: St. Martin's Press, 2015.

McCarthy, John D., Clark McPhail, and Jackie Smith. "Images of Protest: Dimensions of Selection Bias in Media Coverage of Washington Demonstrations, 1982 and 1991." *American Sociological Review* 61, no. 3 (June 1996): 478–99.

McCarthy, John D., and Mayer N. Zald. "Resource Mobilization and Social Movements: A Partial Theory." *American Journal of Sociology* 82, no. 6 (May 1, 1977): 1212–41.

McClearey, Kevin E. "Audience Effects of Apologia." *Communication Quarterly* 31, no. 1 (January 1, 1983): 12–20.

McCormick, Gordon H. "Terrorist Decision Making." *Annual Review of Political Science* 6, no. 1 (June 2003): 473–507.

McCormick, Gordon H., Lindsay Fritz, and Max Abrahms. "Is Suicide Terrorism an Effective Tactic?" In *Debating Terrorism and Counterterrorism: Conflicting Perspectives on Causes, Contexts, and Responses*, edited by Stuart Gottlieb, 2nd ed., 136–71. Washington: CQ Press, 2010.

McCoy, Terrence. "ISIS, Beheadings and the Success of Horrifying Violence." *Washington Post*, June 13, 2014. Morning Mix. https://www.washingtonpost.com/news/morning-mix/wp/2014/06/13/isis-beheadings-and-the-success-of-horrifying-violence/.

McCoy, Terrence. "The Calculated Madness of the Islamic State's Horrifying Brutality." *Washington Post*, August 12, 2014. Morning Mix. https://www.washingtonpost.com/news/morning-mix/wp/2014/08/12/the-calculated-madness-of-the-islamic-states-horrifying-brutality/.

McDermott, Rose. "Experimental Methodology in Political Science." *Political Analysis* 10, no. 4 (2002): 325–42.

McFaul, Michael. "One Step Forward, Two Steps Back." *Journal of Democracy* 11, no. 3 (July 2000): 19–33.

McFaul, Michael. "U.S. Foreign Policy and Chechnya." Washington, DC: Twentieth Century Foundation, 2003. https://cisac.fsi.stanford.edu/sites/default/files/US_Foreign_Policy_and_Chechnya.pdf.

McGreal, Chris. "Arafat Calls off Palestinian Elections." *The Guardian*, December 23, 2002. World news. http://www.theguardian.com/world/2002/dec/23/israel.

McKay, Al. "Interview - Max Abrahms." *E-International Relations*, July 12, 2015. http://www.e-ir.info/2015/07/12/interview-max-abrahms/.

McKernan, Bethan. "Iraq Just Declared Isis's Caliphate 'Has Fallen' after Huge Symbolic Victory in Mosul." *The Independent*, June 29, 2017. Middle East. http://www.independent.co.uk/news/world/middle-east/isis-caliphate-destroyed-islamic-state-iraq-northern-mosul-raqqa-us-led-coalition-forces-a7814281.html.

McKinney, Seamus. "Birth and Rise of the IRA—the Real IRA." *Irish News*. December 21, 2007. http://www.nuzhound.com/articles/irish_news/arts2007/dec20_rise_RIRA.php.

McKnight, Gerald. *The Mind of the Terrorist*. 1st ed. London: Michael Joseph, 1974.

Mearsheimer, John J. *The Tragedy of Great Power Politics*. 1st ed. New York: W. W. Norton & Company, 2001.

Mercer, Dennis, and Ginger Mercer. *From Shantytown to Forest: The Story of Norman Duka*. Oakland, Cal: LSM Information Center, 1977.

Merkl, Peter H. "Approaches to the Study of Political Violence." In *Political Violence and Terror: Motifs and Motivations*, edited by Peter H. Merkl, 1st ed., 19–51. Oakland, CA: University of California Press, 1986.

Michaelis, Arno. "Reflections of a Former White Supremacist." *Al Jazeera*, August 28, 2015. http://www.aljazeera.com/indepth/features/2015/08/reflections-white-suprema cist-150828100415193.html.

Michaels, Jim. "Taliban Leader's Grip on Insurgency Weakens." *USA Today*, January 24, 2012. World. http://www.usatoday.com/news/world/afghanistan/story/2012-01-23/tali ban-leader-afghanistan-war/52760582/1.

Michel, Lou, and Dan Herbeck. *American Terrorist: Timothy McVeigh & the Oklahoma City Bombing*. 1st ed. New York: Harper, 2001.

"Middle East and North Africa." *Military Balance* 103, no. 1 (October 2003): 90–125.

"Middle East/Afghanistan: Al Qaida Threat,." Policy Brief. Oxford Analytica, May 2002. http://www.ciaonet.org/record/14075?search=1.

Miglani, Sanjeev. "Targeted Killings inside Pakistan—Are They Working?" *Reuters Blogs*, August 11, 2009. http://blogs.reuters.com/pakistan/2009/08/12/targeted-killings-inside-pakistanare-they-working/.

Miks, Jason. "Will ISIS Brutality Backfire?" *CNN*, September 26, 2014. http://us.cnn.com/2014/09/25/opinion/cruickshank-storer-hamid-bakos-isis/index.html?sr=sharebar_twitter.

Military Operations in Low Intensity Conflict. Washington, DC: Departments of the Army and Air Force, 1990. https://prariebrand.files.wordpress.com/2015/10/low-intesity-conflict-imp.pdf.

Miller, Stuart. "'We're at War and if that Means More Bombs, So Be It...'" *The Guardian*, April 26, 1999. Global. http://www.theguardian.com/theguardian/1999/apr/27/features11.g2.

Mintzberg, Henry. "The Structuring of Organizations." In *Readings in Strategic Management*, 322–52. London: Macmillan Publishers, 1989. doi:10.1007/978-1-349-20317-8_23.

Moghadam, Assaf. "Palestinian Suicide Terrorism in the Second Intifada: Motivations and Organizational Aspects." *Studies in Conflict & Terrorism* 26, no. 2 (March 1, 2003): 65–92.

Moghadam, Assaf. "Suicide Terrorism, Occupation, and the Globalization of Martyrdom: A Critique of Dying to Win." *Studies in Conflict & Terrorism* 29, no. 8 (November 23, 2006): 707–29.

Mohamed, Hussein, Eric Schmitt, and Mohamed Ibrahim, "Mogadishu Truck Bombings Are Deadliest Attack in Decades." *New York Times*, October 15, 2017. https://www.nytimes.com/2017/10/15/world/africa/somalia-bombing-mogadishu.html.

Moloney, ed. *Voices from the Grave: Two Men's War in Ireland*. 1st ed. New York: PublicAffairs, 2010.

"Monthly US Drone Strikes Quadrupled Under Trump: Report." *TeleSUR*, June 13, 2017. World. https://www.telesurtv.net/english/news/Monthly-US-Drone-Strikes-Quadrupled-Under-Trump-Report-20170613-0034.html.

Moriarty, Gerry. "IRA Dissidents Are Suspected of Being behind Car-Bomb Blast in Banbridge." *The Irish Times*, August 3, 1998. News. https://www.irishtimes.com/news/ira-dissidents-are-suspected-of-being-behind-car-bomb-blast-in-banbridge-1.179463.

Moritz, Cyndi. "ISIS Terrorism: A Q&A with Faculty Experts." *SU News*, November 20, 2015. https://news.syr.edu/2015/11/isis-terrorism-a-qa-with-faculty-experts-33286/.

Morris, Aldon D., and Suzanne Staggenborg. "Leadership in Social Movements." In *The Blackwell Companion to Social Movements*, edited by David A. Snow, Sarah A. Soule, and Hanspeter Kriesi, 171–96. Blackwell Publishing Ltd, 2004. doi:10.1002/9780470999103.ch8.

Morris, Chris. *Four Lions*, 2010. http://www.magnetreleasing.com/fourlions/.

Morris, Loveday. "Is This the High School Report Card of the Head of the Islamic State?" *Washington Post*. February 19, 2015. WorldViews. https://www.washingtonpost.com/news/worldviews/wp/2015/02/19/is-this-the-high-school-report-card-of-the-head-of-the-islamic-state/.

Mortimer, Caroline. "Saudi Arabia 'Ready to Send Ground Troops' to Fight Isis in Syria If US-Led Coalition Agrees." *The Independent*, February 4, 2016. Middle East. http://www.independent.co.uk/news/world/middle-east/saudia-arabia-read-to-send-ground-troops-to-fight-isis-in-syria-if-us-led-coalition-agrees-a6854046.html.

Moss, David. *The Politics of Left-Wing Violence in Italy, 1969–85*. New York: Palgrave Macmillan, 1989.

MSNBC News. "Suicide Blast Slays 40 at Afghan Wedding Party." *NBC News*, June 10, 2010. Afghanistan. http://www.nbcnews.com/id/37589087/ns/world_news-south_and_central_asia/t/suicide-blast-slays-afghan-wedding-party/.

Mueller, John. *Overblown: How Politicians and the Terrorism Industry Inflate National Security Threats, and Why We Believe Them*. New York: Free Press, 2006. http://www.myilibrary.com?id=894261.

Mueller, John E., and Mark G. Stewart. *Chasing Ghosts: The Policing of Terrorism*. Oxford, UK: Oxford University Press, 2016.

Mullen, Brian, and Carolyn Copper. "The Relation between Group Cohesiveness and Performance: An Integration." *Psychological Bulletin* 115, no. 2 (1994): 210–27.

Naadem, Bashir Ahmad. "Dead Bodies of 27 Killed by Taliban Recovered." *Pajhwok Afghan News*, March 16, 2015. http://archive.pajhwok.com/en/2008/10/19/dead-bodies-27-killed-taliban-recovered.

Naji, Abu Bakr. *The Management of Suvagery: The Most Critical Stage Through Which the Umma Will Pass*. Translated by William McCants. John M. Olin Institute for Strategic Studies at Harvard University, 2006. https://azelin.files.wordpress.com/2010/08/abu-bakr-naji-the-management-of-savagery-the-most-critical-stage-through-which-the-umma-will-pass.pdf.

Nance, Malcolm. *Defeating ISIS: Who They Are, How They Fight, What They Believe*. New York: Skyhorse Publishing, 2016.

Nasiri, Omar. *Inside the Jihad: My Life with Al Qaeda*. Reprint ed. New York: Basic Books, 2008.

National Commission on Terrorist Attacks, Thomas H. Kean, and Lee H. Hamilton. *The 9/11 Report: The National Commission on Terrorist Attacks Upon the United States*. New York: St. Martin's Press, 2004.

National Consortium for the Study of Terrorism and Responses to Terrorism (START). 2003. Global Terrorism Database: Al-Aqsa Martyrs Brigades. College Park: University of Maryland. http://www.start.umd.edu/gtd. Accessed 25 June 2014.

Nepstad, Sharon, and Bob Clifford. "When Do Leaders Matter? Hypotheses on Leadership Dynamics in Social Movements." *Mobilization* 11, no. 1 (February 2006): 1–22. doi:https://doi.org/10.17813/maiq.11.1.013313600164m727.

Neumann, Peter. "Victims, Perpetrators, Assets: The Narratives of Islamic State Defectors." The International Cenre for the Study of Radicalisation and Political Violence, 2015. http://icsr.info/wp-content/uploads/2015/09/ICSR-Report-Victims-Perpertrators-Assets-The-Narratives-of-Islamic-State-Defectors.pdf.

Nolan, Cathal J. "'Bodyguard of Lies': Franklin D. Roosevelt and Defensible Deceit in World War II." In *Ethics and Statecraft: The Moral Dimension of International Affairs, 3rd Edition: The Moral Dimension of International Affairs*, edited by Cathal J. Nolan, 3rd ed., 3–29. Boulder, CO: Praeger, 2015.

Norton, Augustus Richard. *Hezbollah: A Short History*. Princeton Studies in Muslim Politics. Princeton: Princeton University Press, 2007.

Nye, Joseph S. Jr. *The Power to Lead: Soft, Hard and Smart Power*. Oxford, UK: Oxford University Press, 2008.

O'Day, Alan. *Political Violence in Northern Ireland: Conflict and Conflict Resolution*. Westport, CT: Praeger, 1997.

Ohbuchi, K., M. Kameda, and N. Agarie. "Apology as Aggression Control: Its Role in Mediating Appraisal of and Response to Harm." *Journal of Personality and Social Psychology* 56, no. 2 (February 1989): 219–27.

O'Keefe, Ed, and Avi Selk. "After @tedcruz Liked a Porn Tweet, Sen. Ted Cruz Blamed It on a Staffer's 'Honest Mistake.'" *Washington Post*, September 13, 2017. Power Post. https://www.washingtonpost.com/news/powerpost/wp/2017/09/12/after-tedcruz-liked-a-porn-tweet-sen-ted-cruz-blamed-a-staffing-issue/.

Olive, Noemie. "Islamists Attack French Church, Slit Priest's Throat." *Reuters*, July 26, 2016. https://www.reuters.com/article/us-france-hostages/two-knifemen-take-several-hostages-in-french-church-police-source-idUSKCN1060VA.

Olsen, Mancur. "The Logic of Collective Action in Soviet-Type Societies." *Journal of Soviet Nationalities* 1, no. 2 (1990): 8–28.

O'Malley, JP. "The West Fundamentally Misunderstands the Ethos of the 'Caliphate,' Writes Muslim Author." *The Times of Israel*, July 6, 2016. http://www.timesofisrael.com/two-state-solution-wont-solve-mideast-crisis-says-brookings-expert/.

O'Malley, Padraig. *Shades of Difference: Mac Maharaj and the Struggle for South Africa*. New York: Viking Adult, 2007.

Omar, Mullah Mohammad. "Message of Felicitation of Amir-Ul-Momineen on the Occasion of Eid-Ul-Fitr." *Voice of Jihad*, August 6, 2013. http://www.shahamat-english.com/index.php/paighamoona/35234-message-of-felicitation-of-amir-ul-momineen-may-allah-protect-him-on-the-occasion-of-eid-ul-fitr.

"Opinion Polls." *Russian Presidential Elections 1996*. Accessed October 7, 2017. http://www.cs.ccsu.edu/~gusev/russian/polls.html.

O'Rawe, Richard. *Blanketmen: An Untold Story of the H-Block Hunger Strike*. Dublin, Ireland: New Island, 2016.

"Osama Bin Laden Speech Offers Peace Treaty with Europe, Says Al-Qa'ida 'Will Persist in Fighting' the U.S." Special Dispatch. The Middle East Media Research Institute, April 15, 2004. https://www.memri.org/reports/osama-bin-laden-speech-offers-peace-treaty-europe-says-al-qaida-will-persist-fighting-us

Osborn, Andrew. "Putin Vows Payback after Confirmation of Egypt Plane Bomb." *Reuters*, November 17, 2015. World News. https://www.reuters.com/article/us-egypt-crash-russia-blast/russia-says-bomb-did-down-airbus-a321-plane-over-egypt-last-month-idUSKCN0T60PS20151117.

Osborne, Samuel. "Syria Air Strike: At Least 42 Civilians Killed by US-Led Bomb Attack in Raqqa, Report Activists." *The Independent*, August 22, 2017. http://www.independent.co.uk/news/world/middle-east/raqqa-syria-air-strike-us-coalition-civilians-killed-isis-sdf-assad-observatory-a7905966.html?amp.

Osiel, Mark J., ed. *Mass Atrocity, Collective Memory, and the Law*. London: Routledge, 1999.

Ostrom, Elinor. *Governing the Commons: The Evolution of Institutions for Collective Action*. 1st edition. Cambridge, UK: Cambridge University Press, 1990.

Overgaard, Per Baltzer. "The Scale of Terrorist Attacks as a Signal of Resources." *Journal of Conflict Resolution* 38, no. 3 (September 1, 1994): 452–78.

Pape, Robert A. "The Strategic Logic of Suicide Terrorism." *The American Political Science Review* 97, no. 3 (2003): 343–61.

Pape, Robert A. *Dying to Win: The Strategic Logic of Suicide Terrorism*. 1st ed. New York: Random House, 2005.

Pape, Robert A., Jenna McDermit, and Lindsey O'Rourke. "What Makes Chechen Women So Dangerous?" *The New York Times*, March 30, 2010. Opinion. https://www.nytimes.com/2010/03/31/opinion/31pape.html.

"Partidos Próximos a Las Dos Ramas de ETA Condenan Los Atentados." *EL PAÍS*, August 1, 1979.

"Partners." *The Global Coalition Against Daesh*. Accessed October 5, 2017. http://theglobalcoalition.org/en/partners/.

Pearlman, Wendy. *Violence, Nonviolence, and the Palestinian National Movement*. Cambridge, UK: Cambridge University Press, 2011.

Pearlman, Wendy. *Violence, Nonviolence, and the Palestinian National Movement*. 1st ed. Cambridge, UK: Cambridge University Press, 2011.

Pearson, Frederic S., Isil Akbulut, and Marie Olson Lounsbery. "Group Structure and Intergroup Relations in Global Terror Networks: Further Explorations." *Terrorism and Political Violence* 29, no. 3 (May 4, 2017): 550–72.

Pearson, Michael, Greg Botelho, and Ben Brumfield. "Anti-ISIS Coalition Grows, but Challenges Remain." *CNN*. World. Accessed October 5, 2017. http://www.cnn.com/2014/09/26/world/meast/isis-syria-iraq/index.html.

Pedahzur, Ami. *Suicide Terrorism*. 1st ed. Cambridge, UK: Polity, 2005.

Perritano, John. "10 Scapegoats Still in the Crosshairs." *How Stuff Works*, April 24, 2013. http://history.howstuffworks.com/historical-figures/10-scapegoats-still-in-crosshairs.htm.

"Philippine Rebels Apologize for Attack on Medics." *Ucanews.Com*. March 7, 2014. https://www.ucanews.com/news/philippine-rebels-apologize-for-attack-on-medics/70446.

"PM 'Says Israel Pre-Planned War.'" *BBC News*, March 8, 2007. http://news.bbc.co.uk/2/hi/middle_east/6431637.stm.

Pool, Patrick. "U.S. Warns the World About 19,000 ISIS Fighters on the Loose." *PJ Media*, July 27, 2017. https://pjmedia.com/homeland-security/2017/07/27/breaking-u-s-warns-the-world-about-19000-isis-fighters-on-the-loose/.

Post, Jerrold, Ehud Sprinzak, and Laurita Denny. "The Terrorists in Their Own Words: Interviews with 35 Incarcerated Middle Eastern Terrorists." *Terrorism and Political Violence* 15, no. 1 (March 2003): 171–84.

Price, Bryan C. "Targeting Top Terrorists: How Leadership Decapitation Contributes to Counterterrorism." *International Security* 36, no. 4 (April 1, 2012): 9–46.

Price, Robert M. *The Apartheid State in Crisis: Political Transformation in South Africa, 1975–1990*. 1st ed. Oxford, UK: Oxford University Press, 1991.

Primoratz, Igor, ed. *Civilian Immunity in War*. 1st ed. New York: Clarendon Press, 2007.

Prudori, Tony. "On Civilian Casualties, Taliban Try to Look like Victims." *Threat Matrix: FDD's Long War Journal*, August 17, 2010. https://www.longwarjournal.org/archives/2010/08/on_civilian_casualties_taliban_1.php.

Public Opinion Foundation. "The Bombing of Houses in Russian Towns." Фонд Общественное Мнение, September 14, 2000. http://bd.english.fom.ru/report/cat/az/A/explosion_house/eof003701.

Public Opinion Foundation. "Chechen Labyrinth." Фонд Общественное Мнение, June 6, 2002. http://bd.english.fom.ru/report/cat/az/A/chechenian/ed022239.

Purkis, Jonathan. "Leaderless Cultures: The Problem of Authority in a Radical Environmental Group." In *Leadership and Social Movements*, edited by Colin Barker, Alan Johnson, and Michael Lavalette. Manchester, UK: Manchester University Press, 2001.

"Putin Meets Angry Beslan Mothers." *BBC News*, September 2, 2005. http://news.bbc.co.uk/2/hi/europe/4207112.stm.

Pye, Lucian W. *Guerrilla Communism in Malaya: Its Social and Political Meaning*. Princeton: Princeton University Press, 1956.

Qassem, Naim. *Hizbullah: The Story from Within*. London: Saqi Books, 2010.

Quinn, Patrick, and Katherine Shrader. "Foreigners Responsible for Most Suicide Attacks in Iraq." *Lincoln Journal Star*. June 30, 2005. National News. http://journalstar.com/news/national/foreigners-responsible-for-most-suicide-attacks-in-iraq/article_a5096f1f-1d4d-56da-a312-13c095ec5214.html.

Raines, Howell. "Terrorism; With Latest Bomb, I.R.A. Injures Its Own Cause." *The New York Times*. November 15, 1987. World. http://www.nytimes.com/1987/11/15/weekinreview/the-world-terrorism-with-latest-bomb-ira-injures-its-own-cause.html.

Ranstorp, Magnus. *Hizb'allah in Lebanon: The Politics of the Western Hostage Crisis*. 1st ed. London: Palgrave Macmillan UK, 1997.

Raphaeli, Nimrod. " 'The Sheikh of the Slaughterers': Abu Mus'ab Al-Zarqawi and the Al-Qaeda Connection." Inquiry & Analysis. The Middle East Media Research Institute, June 30, 2005. https://www.memri.org/reports/%E2%80%98-sheikh-slaughterers%E2%80%99-abu-musab-al-zarqawi-and-al-qaeda-connection.

Rapoport, David C. "The Four Waves of Rebel Terror and September 11." *Anthropoetics* 8, no. 1 (2002). http://anthropoetics.ucla.edu/ap0801/terror/.

Rassler, Don, Gabriel Koehler-Derrick, Liam Collins, Muhammad al-Obaidi, and Nelly Lahoud. "Letters from Abbottabad: Bin Ladin Sidelined?" HARMONY PROGRAM. West Point, NY: Combating Terrorism Center, May 2, 2012. https://ctc.usma.edu/posts/letters-from-abbottabad-bin-ladin-sidelined.

"Rebel Leader Apologizes for Colombian Church Kidnapping." *CNN*, June 7, 1999. Americas. http://www.cnn.com/WORLD/americas/9906/07/colombia/.

Reiter, Dan. *Crucible of Beliefs: Learning, Alliances, and World Wars*. Cornell Studies in Security Affairs. Ithaca, NY: Cornell University Press, 1996.

Rej, Abhijnan. "The Strategist: How Abu Mus'ab Al-Suri Inspired ISIS." Occasional Paper. Observer Research Foundation, August 2016. http://www.orfonline.org/research/the-strategist-how-abu-musab-al-suri-inspired-isis/.

"Report: PKK Punished Those Behind Batman Blast." *Today's Zaman*, December 2, 2010. http://www.todayszaman.com/national_report-pkk-punished-those-behind-batman-blast-apologized-for-deaths_228438.html.

Reuters Staff. "Al Qaeda-Linked Militants in Lebanon Apologize for Civilian Deaths." *Reuters*, March 8, 2014. World News. https://www.reuters.com/article/us-lebanon-blasts/al-qaeda-linked-militants-in-lebanon-apologize-for-civilian-deaths-idUSBREA270L120140308.

Reuters Staff. "Fewer Foreign Fighters Joining Islamic State: Pentagon." *Reuters*, April 26, 2016. World News. https://www.reuters.com/article/us-mideast-crisis-recruiting/fewer-foreign-fighters-joining-islamic-state-pentagon-idUSKCN0XN2FO.

Reuters Staff. "Captured ISIS Militant Says He Raped and Killed Hundreds: 'It Was Normal.'" *New York Post*, February 17, 2017. http://nypost.com/2017/02/17/captured-isis-militant-says-he-raped-and-killed-hundreds-it-was-normal/.

Reuters Staff. "Islamic State 'Bureaucrats' Fleeing Stronghold in Syria: Pentagon." *Reuters*, February 17, 2017. World News. https://www.reuters.com/article/us-mideast-crisis-raqqa-pentagon/islamic-state-bureaucrats-fleeing-stronghold-in-syria-pentagon-idUSKBN15W24E.

Reuters Staff. "Islamic State Has Lost Most Territory It Held in Iraq: Iraqi Spokesman." *Reuters*, April 11, 2017. World News. https://www.reuters.com/article/us-mideast-crisis-iraq-mosul/islamic-state-has-lost-most-territory-it-held-in-iraq-iraqi-spokesman-idUSKBN17D1FP.

Reuters Staff, and Armin Rosen. "ISIS' Territory in Syria and Iraq Is Shrinking—but there's a Catch." *Business Insider*, January 5, 2016. http://www.businessinsider.com/r-islamic-state-territory-shrinks-in-iraq-and-syria-us-led-coalition-2016-1.

"Revolutionary Council and DFLP Call for an End to Attacks inside Israel." *Al-Hayat*, May 30, 2002.

Reynolds, Paul D., and Sammis B. White. *The Entrepreneurial Process: Economic Growth, Men, Women, and Minorities*. Westport, CT: Quorum Books, 1997.

Richardson, Andrew. "Sports Psychology - Group Cohesion." *Irish Drug Free Powerlifting Association*, July 23, 2013. http://www.idfpa.net/sports-psychology—group-cohesion.

Richardson, Louise. *What Terrorists Want: Understanding the Enemy, Containing the Threat*. New York: Random House Trade Paperbacks, 2007.

Rigo, Enric Ordeix i. "Aznar's Political Failure or Punishment for Supporting the Iraq War? Hypotheses About the Causes of the 2004 Spanish Election Results." *American Behavioral Scientist* 49, no. 4 (December 2005): 610–15.

Riordan, Catherine A., Nancy A. Marlin, and Catherine Gidwani. "Accounts Offered for Unethical Research Practices: Effects on the Evaluations of Acts and Actors." *The Journal of Social Psychology* 128, no. 4 (August 1, 1988): 495–505.

Rivoire, Jean-Baptiste, and Jean-Paul Billault. "Bentalha: Autopsie d'un Massacre." *Algeria-Watch*, September 23, 1999. http://www.algeria-watch.org/farticle/bentalha/Bentalhafilm.htm.

"Robert Gates: Expect More Attacks in the West as ISIS Is Defeated in the Middle East." *Washington Times*. May 23, 2017. http://www.washingtontimes.com/news/2017/may/23/robert-gates-expect-more-attacks-west-isis-defeate/.

Roberts, Dan. "Hillary Clinton: Trump's Rhetoric about Terrorist Attacks 'Gives Isis What It Wants.'" *The Guardian*, September 19, 2016. US News. http://www.theguardian.com/us-news/2016/sep/19/hillary-clinton-new-york-new-jersey-attacks-trump-muslims.

Robertson, Geoffrey. *Crimes Against Humanity: The Struggle for Global Justice by Geoffrey Robertson*. New York: The New Press, 1999.

Rochon, Thomas R. *Between Society and State: Mobilizing for Peace in Western Europe*. Princeton: Princeton University Press, 1988.

Roggio, Bill. "Taliban Suicide Bomber Kills 32 in Afghan North." *FDD's Long War Journal*, February 21, 2011. https://www.longwarjournal.org/archives/2011/02/taliban_suicide_bomb_20.php.

Roggio, Bill. "Taliban Kill 21 Afghans in Double Suicide Attack." *FDD's Long War Journal*, June 6, 2012. https://www.longwarjournal.org/archives/2012/06/taliban_killed_21_af.php.

Roggio, Bob. "Suicide Bomber Strikes in Western Afghanistan." *FDD's Long War Journal*, November 20, 2009. https://www.longwarjournal.org/archives/2009/11/suicide_bomber_strik_3.php.

Roggio, Bob, and Bill Ardolino. "New Details Emerge about Complex Attack on FOB Salerno." *FDD's Long War Journal*, June 10, 2012. https://www.longwarjournal.org/archives/2012/06/new_details_emerge_a.php.

Romano, Lois. "Prosecutors Seek Death For Nichols." *Washington Post*. December 30, 1997.

Rose, William, Rysia Murphy, and Max Abrahms. "Does Terrorism Ever Work? The 2004 Madrid Train Bombings." *International Security* 32, no. 1 (July 2007): 185–92.

Ross, Jeffrey Ian. "Structural Causes of Oppositional Political Terrorism: Towards a Causal Model." *Journal of Peace Research* 30, no. 3 (August 1, 1993): 317–29.

Rothwell, James, Chris Graham, and Barney Henderson. "German Axe Attack on Train: Isil Claim Afghan Refugee Who Injured Four as One of Its 'Fighters.'" *The Telegraph*, July 18, 2016. News. http://www.telegraph.co.uk/news/2016/07/18/german-train-axe-attack-many-injured/.

Roy, Beth, John Burdick, and Louis Kriesberg. "A Conversation between Conflict Resolution and Social Movement Scholars." *Conflict Resolution Quarterly* 27, no. 4 (July 9, 2010): 347–68.

Russell, Cameron Charles. "The Targeted Killings Debate – the Endogeneity Problem." *Across the Pond*, June 14, 2011. https://irpond.wordpress.com/2011/06/14/the-targeted-killings-debate-the-endogeneity-problem/.

Russell, John. "Mujahedeen, Mafia, Madmen: Russian Perceptions of Chechens During the Wars in Chechnya, 1994-96 and 1999-2001." *Journal of Communist Studies and Transition Politics* 18, no. 1 (March 1, 2002): 73–96.

Russell, John. "Terrorists, Bandits, Spooks and Thieves: Russian Demonization of the Chechens before and since 9/11." *Third World Quarterly* 26, no. 1 (February 2005): 101–16.

"Russia Intensifies Airstrikes on IS-Stronghold Deir Az-Zour." *The New Arab*, August 22, 2017. https://www.alaraby.co.uk/english/news/2017/8/22/russia-intensifies-airstrikes-on-is-stronghold-deir-az-zour.

Ryan, Halford Ross. "Prime Minister Stanley Baldwin vs King Edward VIII." In *Oratorical Encounters: Selected Studies and Sources of Twentieth-Century Political Accusations and Apologies*. Contributions to the Study of Mass Media and Communications, no. 9. New York: Greenwood Press, 1988.

Sageman, Marc. *Understanding Terror Networks*. Philadelphia: University of Pennsylvania Press, 2004.

Sageman, Marc. *Leaderless Jihad: Terror Networks in the Twenty-First Century*. Philadelphia: University of Pennsylvania Press, 2008.

Salem, Arab, Edna Reid, and Hsinchun Chen. "Multimedia Content Coding and Analysis: Unraveling the Content of *Jihadi* Extremist Groups' Videos." *Studies in Conflict & Terrorism* 31, no. 7 (June 24, 2008): 605–26.

Sanday, Peggy Reeves. *Fraternity Gang Rape: Sex, Brotherhood, and Privilege on Campus*. New York: NYU Press, 1990.

Sandler, Todd, and John L. Scott. "Terrorist Success in Hostage-Taking Incidents: An Empirical Study." *The Journal of Conflict Resolution* 31, no. 1 (1987): 35–53.

Satter, David. "The Mystery of Russia's 1999 Apartment Bombings Lingers—the CIA Could Clear It Up." *National Review*, February 2, 2017. http://www.nationalreview.com/article/444493/cia-russias-1999-apartment-bombings-if-putin-was-responsible-it-could-well-know.

Saunders, Elizabeth N. *Leaders at War: How Presidents Shape Military Interventions*. Cornell Studies in Security Affairs. Ithaca, NY: Cornell University Press, 2011.

Savage, Charlie, and Eric Schmitt. "Trump Administration Is Said to Be Working to Loosen Counterterrorism Rules." *The New York Times*, March 12, 2017. Politics. https://www.nytimes.com/2017/03/12/us/politics/trump-loosen-counterterrorism-rules.html?nytmobile=0.

Sayigh, Yezid. "Arafat and the Anatomy of a Revolt." *Survival* 43, no. 3 (September 2001): 47–60.

Schelling, Thomas C. "What Purposes Can 'International Terrorism' Serve?" In *Violence, Terrorism, and Justice*, edited by Raymond Gillespie Frey and Christopher W. Morris, 18–32. Cambridge, UK: Cambridge University Press, 1991. doi:10.1017/CBO9780511625039.003.

Schelling, Thomas C. *Arms and Influence*. New Haven: Yale University Press, 1966.

Scheuer, Michael. *Imperial Hubris: Why the West Is Losing the War on Terror*. Dulles, VA: Potomac Books, Inc, 2004.

Scheuer, Michael. *Osama Bin Laden*. 1st ed. Oxford UK: Oxford University Press, 2011.

Schlenker, Barry R. *Impression Management: The Self-Concept, Social Identity, and Interpersonal Relations*. Monterey, CA: Brooks/Cole, 1980.

Schmid, Alex P. "Why Terrorism? Root Causes, Some Empirical Findings, and the Case of 9/11." Speech presented at the Council of Europe, Strasbourg, France, April 26, 2007.

Schmid, Alex P., and Janny de Graaf. *Violence as Communication: Insurgent Terrorism and the Western News Media*. Beverly Hills, CA: SAGE Publications, 1982.

Schmid, Alex P., and Albert J. Jongman. "Political Terrorism: A New Guide to Actors and Authors." In Schmid, A.P., ed. *Data Bases, and Literature,* New Brunswick: Transaction Books (2005).

Schmitt, Eric. "As Al Qaeda Loses a Leader, Its Power Shifts From Pakistan." *The New York Times,* June 7, 2012. Asia Pacific. https://www.nytimes.com/2012/06/08/world/asia/al-qaeda-power-shifting-away-from-pakistan.html.

Schmitt, Khaled. "Russia Killed 1123 Chechen Civilians in 2002: Report." *IslamOnline,* April 23, 2004. https://worldofislam.info/news/index.php?page=Chechnya/Russia%20Killed%201123%20Chechen%20Civilians%20in%202002%20Report.

Schreck, Adam, and Zeina Karam. "A Look At The Dangers Posed By The Islamic State; Brutal Efficiency, The Key To The Islamic States Success." *Fortuna's Corner,* September 2, 2014. https://fortunascorner.com/2014/09/03/a-look-at-the-dangers-posed-by-the-islamic-state-brutal-efficiency-the-key-to-the-islamic-states-success/.

Schultz, Kenneth A. "Looking for Audience Costs." *The Journal of Conflict Resolution* 45, no. 1 (2001): 32–60.

Schutte, Shane. "CEOs That Used Staff as Scapegoats to Protect the Firm—or Themselves." *Real Business,* March 9, 2016. https://realbusiness.co.uk/hr-and-management/2016/03/09/ceos-that-used-staff-as-scapegoats-to-protect-the-firm-or-themselves/.

Scott, Marvin B., and Stanford M. Lyman. "Accounts." *American Sociological Review* 33, no. 1 (February 1968): 46–62.

Scott, W. Richard. "Developments in Organization Theory, 1960–1980." *American Behavioral Scientist* 24, no. 3 (January 1, 1981): 407–22.

"Self-Styled Muslim Sheikh on Accessory to Murder Charge." *ABC News,* November 22, 2013. http://www.abc.net.au/news/2013-11-22/self-styled-muslim-sheikh-faces-court-on-accessory-to-murder-ch/5111744.

Shah, Aqil. "Democracy on Hold in Pakistan." *Journal of Democracy* 13, no. 1 (2002): 67–75.

Shahzad, Asif, and Kimberly Dozier. "Record Level of US Airstrikes Hit Afghan Militants." *Boston.Com,* September 14, 2010.

Shapiro, Jacob N. *The Terrorist's Dilemma: Managing Violent Covert Organizations*. Princeton: Princeton University Press, 2013.

Sharkov, Damien. "ISIS Video Shows Training of Child Soldiers." *Newsweek,* December 10, 2014. World. http://www.newsweek.com/isis-video-shows-training-child-soldiers-290768.

Sharp, Gene. "The Intifadah and Nonviolent Struggle." *Journal of Palestine Studies* 19, no. 1 (October 1989): 3–13.

Sharp, Gene, and Marina Finkelstein. *The Dynamics of Nonviolent Action*. The Politics of Nonviolent Action. New York: Porter Sargent, 1973.

Shevtsova, Lilia. *Yeltsin's Russia: Myths and Reality*. Washington, DC: Carnegie Endowment for International Peace, 1999.

Shiloach, Gilad. "Two Prominent ISIS Commanders Reportedly Defect." *Vocativ*, December 6, 2016. http://www.vocativ.com/381169/two-prominent-isis-commanders-reportedly-defect/.

Shlaim, Avi. "When Bush Comes to Shove: America and the Arab-Israeli Peace Process." *Oxford International Review* 3, no. 2 (Spring 1992): 2–6.

Siegel, Jacob. "Islamic Extremists Now Crucifying People in Syria—and Tweeting Out the Pictures." *The Daily Beast*, April 30, 2014. World. https://www.thedailybeast.com/articles/2014/04/30/islamic-extremists-now-crucifying-people-in-syria-and-tweeting-out-the-pictures.

Sifaoui, Mohamed. *Inside Al Qaeda: How I Infiltrated the World's Deadliest Terrorist Organization*. New York: Thunder's Mouth Press, 2004.

Sinclair, Andrew. *Che Guevara*. New York: Viking Press, 1970.

Singh, Tavleen. "India's Sikh Extremists Scramble to Recoup Mainstream Support. Sikh Terror Campaign Seen as Effort to Provoke Hindu Backlash." *Christian Science Monitor*, June 23, 1988. https://www.csmonitor.com/1988/0623/ojab.html.

SITE Staff. "Abu Musab Al-Suri's Military Theory of Jihad." *Abu Musab Al-Suri's Military Theory of Jihad*, October 14, 2010. http://news.siteintelgroup.com/blog/index.php/about-us/21-jihad/21-suri-a-mili.

Slim, Hugo. *Killing Civilians: Method, Madness and Morality in War*. London: Hurst, 2007.

Slovo, Joe. *Slovo: The Unfinished Autobiography*. Randburg: Ravan Press, 1997.

Smith, Mark A. "The Second Chechen War: The All-Russian Context." Occasional Paper. London: Strategic and Combat Studies Institute, September 2000.

Smith, R. Jeffrey. "A Bosnian Village's Terrorist Ties." *Washington Post*, March 11, 2000. http://www.washingtonpost.com/wp-srv/WPcap/2000-03/11/006r-031100-idx.html.

Snow, Robert L. *The Militia Threat: Terrorists among Us*. New York: Da Capo Press, 1999.

Snyder, Charles R. "The Excuse: An Amazing Grace?" In *The Self and Social Life*, edited by Barry R Schlenker. New York: McGraw-Hill, 1985. http://trove.nla.gov.au/version/26729493.

Snyder, Charles R., and Raymond L. Higgins. "Excuses: Their Effective Role in the Negotiation of Reality." *Psychological Bulletin* 104, no. 1 (July 1988): 23–35.

Soguel, Dominique. "Syrian Druze Massacre: Can Jihadists Salvage Their Image?" *Christian Science Monitor*. Accessed October 6, 2017. https://www.csmonitor.com/World/Middle-East/2015/0615/Syrian-Druze-massacre-Can-jihadists-salvage-their-image.

"Somali 'Killed for Not Wearing Veil.'" *BBC News*, July 30, 2014. Africa. http://www.bbc.com/news/world-africa-28564984.

Sparks, Allister Haddon. *The Mind of South Africa*. New York: Ballantine, 1991.

Spiers, Ronald. "Confusion in America : Try Clearer Thinking about 'Terrorists.'" *The New York Times*. January 14, 2003. Opinion. https://www.nytimes.com/2003/01/14/opinion/confusion-in-america-try-clearer-thinking-about-terrorists.html.

Staniland, Paul. *Networks of Rebellion: Explaining Insurgent Cohesion and Collapse*. Cornell Studies in Security Affairs. Ithaca, NY: Cornell University Press, 2014.

Stanton, Jessica. *Violence and Restraint in Civil War: Civilian Targeting in the Shadow of International Law*. New York: Cambridge University Press, 2016.

Stanton, Jessica. *Rebel Groups, International Humanitarian Law, and Civil War Outcomes in the Post-Cold War Era.* Philadelphia: University of Pennsylvania, 2017.

Starr, Barbara. "Trump Looks for Good News in ISIS Fight Ahead of NATO Summit." *CNN*, May 24, 2017. Politics. http://www.cnn.com/2017/05/24/politics/trump-nato-isis/index.html.

Steel, Ronald. "Blowback: Terrorism and the U.S. Role in the Middle East." *New Republic*, July 28, 1996.

Stern, Jessica. *Terror in the Name of God: Why Religious Militants Kill.* 1st ed. New York: Ecco, 2003.

Stern, Jessica. "The Protean Enemy." *Foreign Affairs*, July 1, 2003. https://www.for eignaffairs.com/articles/afghanistan/2003-07-01/protean-enemy.

Stern, Jessica, and J. M. Berger. *ISIS: The State of Terror.* New York: Harper Collins, 2015.

Stern, Susan. *With the Weathermen: The Personal Journal of a Revolutionary Woman.* New Brunswick, N.J: Rutgers University Press, 2007.

Stone, Martin. *The Agony of Algeria.* New York: Columbia University Press, 1997.

Stork, Joe. *Erased in a Moment: Suicide Bombing Attacks against Israeli Civilians.* New York: Human Rights Watch, 2002. https://www.hrw.org/reports/2002/isrl-pa/ISRAELPA1002.pdf.

Strenger, Carlo. "France Must Fight Terrorism Without Harming The Spirit Of Liberty." *Huffington Post*, July 20, 2016. https://www.huffingtonpost.com/carlo-strenger/france-must-fight-terrori_b_11038360.html.

Strentz, Thomas. "A Terrorist Psychosocial Profile: Past and Present." *FBI Law Enforcement Bulletin*, 1988.

Strick van Linschoten, Alex, and Felix Kuehn. "Separating the Taliban from Al-Qaeda: The Core of Success in Afghanistan." Center on International Cooperation, February 2011. http://cic.es.its.nyu.edu/sites/default/files/gregg_sep_tal_alqaeda.pdf.

Suarez, Ray, José Gijon, and Salvador Sala, "Terrorism and Politics in Spain." *NewsHour*, March 16, 2004.

Subrahmanian, V. S., Aaron Mannes, Amy Sliva, Jana Shakarian, and John P. Dickerson. *Computational Analysis of Terrorist Groups: Lashkar-e-Taiba.* New York: Springer, 2013.

"Suicide Attack in Afghanistan Kills at Least 33 People." *France 24*, April 18, 2015. Middle East. http://www.france24.com/en/20150418-afghanistan-suicide-bombing-jalalabad-bank/.

"Suicide Blast Slays 40 at Afghan Wedding Party." *MSNBC*, June 10, 2010. Afghanistan. http://www.nbcnews.com/id/37589087/ns/world_news-south_and_central_asia/t/suicide-blast-slays-afghan-wedding-party/.

Suskind, Ron. *The Price of Loyalty: George W. Bush, the White House, and the Education of Paul O'Neill.* New York: Simon & Schuster, 2004.

"Sydney Siege Inquest Builds Picture of Man Haron Monis' 'Bizarre, Delusional' World." *7.20 with Leigh Sales*, May 25, 2015. http://www.abc.net.au/7.30/sydney-siege-inquest-builds-picture-of-man-haron/6496402.

Sykes, Gresham M., and David Matza. "Techniques of Neutralization: A Theory of Delinquency." *American Sociological Review* 22, no. 6 (1957): 664–70.

"Syrian Rebels Claim Regime Using Civilians as 'Human Shields.'" *The New Arab*, November 2, 2016. https://www.alaraby.co.uk/english/news/2016/11/2/syrian-rebels-claim-regime-using-civilians-as-human-shields.

"Syria's White Helmets Suspend Members Caught on Camera during Rebel Execution." *RT*, May 20, 2017. https://www.rt.com/news/388960-white-helmets-execution-footage/.

Szakola, Albin. "Captured Hezbollah Members 'Received Little Training.'" *NOW*, September 2, 2016. https://now.mmedia.me/lb/en/NewsReports/566590-captured-hezbollah-members-received-little-training-lost.

Tabbara, Lina Mikdadi. *Survival in Beirut: Diary of Civil War*. English ed. London: Onyx Books, 1979.

Tahmincioglu, Eve. "Should You Accept Blame for a Mistake?" *MSNBC*, October 20, 2008. Careers. http://www.nbcnews.com/id/27239935/ns/business-careers/t/should-you-accept-blame-mistake/.

"Taliban Calls on Fighters to Spare Civilians." *Al Jazeera*, November 6, 2011. http://www.aljazeera.com/news/asia/2011/11/2011115162452523578.html.

"Taliban Propaganda: Winning the War of Words?" Asia. International Crisis Group, July 24, 2008. https://www.crisisgroup.org/asia/south-asia/afghanistan/taliban-propaganda-winning-war-words.

Tanner, Stuart. "Battle for the Holy Land: Interviews with Three Palestinian Militant Leaders." *PBS Frontline*, April 4, 2002.

Tarabay, Jamie, Gilad Shiloach, Amit Weiss, and Matan Gilat. "To Its Citizens, ISIS Shows A Softer Side Too." *Vocativ*, March 20, 2015. http://www.vocativ.com/world/isis-2/to-its-citizens-isis-also-shows-a-softer-side/.

Tarrow, Sidney. *Power in Movement: Social Movements, Collective Action and Mass Politics in the Modern State*. Cambridge, UK: Cambridge University Press, 1994.

Taruc, Luis. *He Who Rides the Tiger: The Story of an Asian Guerrilla Leader*. Westport, CT: Praeger, 1967.

Tawil, Camille. *Brothers In Arms: The Story of Al-Qa'ida and the Arab Jihadists*. Translated by Robin Bray. English ed. London: Saqi Books, 2011.

Taylor, Adam. "Why Would Terrorists Kill Cartoonists?" *Washington Post*, January 7, 2015. WorldViews. https://www.washingtonpost.com/news/worldviews/wp/2015/01/07/why-would-terrorists-kill-cartoonists/.

Taylor, Frederick Winslow. *The Principles of Scientific Management*. New York: Harper, 1914. https//catalog.hathitrust.org/Record/011713389.

Taylor, Peter. *Behind the Mask: The IRA and Sinn Fein*. New York: TV Books, 1999.

Taylor, Peter. "World's Richest Terror Army." *This World*. BBC Two, April 22, 2015. http://www.bbc.co.uk/programmes/b05s4ytp.

Thomas, Timothy. "Manipulating the Mass Consciousness: Russian and Chechen 'Information War' Tactics in the Second Chechen-Russian Conflict." In *The Second Chechen War: The All-Russian Context*, edited by Mark A. Smith. Washington, DC: Conflict Studies Research Center, 2000.

Thrasher, Frederic Milton. *The Gang: A Study of 1,313 Gangs in Chicago*. 2nd ed. University of Chicago Sociological Series. Chicago: The University of Chicago Press, 2013.

Tilly, Charles. *The Politics of Collective Violence*. Cambridge Studies in Contentious Politics. Cambridge, UK: Cambridge University Press, 2003.

TNN. "Worms Creep out of Cadbury." *The Times of India*, October 12, 2003. https://timesofindia.indiatimes.com/india/Worms-creep-out-of-Cadbury/articleshow/229817.cms.

Toolis, Kevin. *Rebel Hearts: Journeys within the IRA's Soul*. New York: St. Martin's Press, 2015. http://rbdigital.oneclickdigital.com.

"Top MP Killed in Suicide Attack at Daughter's Wedding." *France 24*, July 14, 2012. Asia-Pacific. http://www.france24.com/en/20120714-afghanistan-suicide-bomber-kills-22-afghan-wedding-taliban.

Tse-tung, Mao. *On Guerrilla Warfare*. Translated by Samuel B. Griffith. Urbana, IL: University of Illinois Press, 2005.

"Tunisia Passes New Anti-Terror Law." *BBC News*, July 25, 2015. Africa. http://www.bbc.com/news/world-africa-33662633.

Turok, Ben. *Nothing but the Truth: Behind the ANC's Struggle Politics*. Johannesburg: Jonathan Ball Publishers, 2003.

Tversky, Amos, and Daniel Kahneman. "Availability: A Heuristic for Judging Frequency and Probability." *Cognitive Psychology* 5, no. 2 (September 1973): 207–32.

Tynan, Patrick. *The Irish National Invincibles and Their Times*. London: Chatham, 1894.

United Nations Assistance Mission In Afghanistan, and United Nations Office of the High Commissioner for Human Rights. "Protection Of Civilians In Armed Conflict—Afghanistan Annual Report 2013," February 2014. https://unama.unmissions.org/sites/default/files/feb_8_2014_poc-report_2013-full-report-eng.pdf.

"Urgent Appeal to Stop Suicide Bombings." *Al Quds*, June 20, 2002. http://www.bitterlemons.net/docs/suicide.html.

US Department of State. "Joint U.S.—Pakistan Statement for President Musharraf's Visit to New York," November 10, 2001. https://2001-2009.state.gov/p/sca/rls/rm/6060.htm.

U.S. Department of State, "Compilation of Usama Bin Ladin Statements: 1994—January 2004," Foreign Broadcast Information Service (Washington, DC: 2004). http://www.fas.org/irp/world/para/ubl-fbis.pdf.

Usher, Graham. "Facing Defeat: The Intifada Two Years On." *Journal of Palestine Studies* 32, no. 2 (January 2003): 21–40.

Usher, Graham. "Al-Ahram Weekly | Intifada | Still Seeking a Vision." *Al-Ahram Weekly*, September 30, 2004. http://weekly.ahram.org.eg/Archive/2004/710/fo1.htm.

Vallières, Pierre. *White Niggers of America: The Precocious Autobiography of a Quebec "Terrorist."* New York: Monthly Review Press, 1971.

Van Biezen, Ingrid. "Terrorism and Democratic Legitimacy: Conflicting Interpretations of the Spanish Elections." *Mediterranean Politics* 10, no. 1 (March 2005): 99–108.

Vanden Brook, Tom. "Islamic State Defections Mount as Death Toll Rises, U.S. Official Says." *USA TODAY*, November 29, 2015. World. https://www.usatoday.com/story/news/world/2015/11/29/islamic-state-defections-kurds-lloyd-austin-syria-isil/76503736/.

Vargas, Robert. "Criminal Group Embeddedness and the Adverse Effects of Arresting a Gang's Leader: A Comparative Case Study." *Criminology* 52, no. 2 (February 21, 2014): 143–68.

Vasquez, John A. "The Realist Paradigm and Degenerative versus Progressive Research Programs: An Appraisal of Neotraditional Research on Waltz's Balancing Proposition." *American Political Science Review* 91, no. 04 (December 1997): 899–912.

Victor, Barbara. *Army of Roses: Inside the World of Palestinian Women Suicide Bombers*. Harlan, IA: Rodale, 2003.

Vigilant, Lee Garth, and John B. Williamson. "On the Role and Meaning of Death in Terrorism." In *Handbook of Death and Dying*, edited by Clifton D. Bryant, 1:236–45. Thousand Oaks, CA: Sage Publishers, 2003.

Vinograd, Cassandra "Canada Launches First Airstrikes Against ISIS in Iraq." *NBC News*, November 3, 2014. https://www.nbcnews.com/storyline/isis-terror/canada-launches-first-airstrikes-against-isis-iraq-n239741.

Vision Reporter. "Heed Final Call, LRA." *New Vision*, October 10, 2003. http://www.newvision.co.ug/new_vision/news/1258572/heed-final-lra.

Wagner, Daniel. "The Dark Side of the Free Syrian Army." *Huffington Post*, December 31, 2012. https://www.huffingtonpost.com/daniel-wagner/dark-side-free-syrian_b_2380399.html.

Wahedi, Laila A. "Anti-Social Networks: The Effects of Violent Group Cooperative Networks Structure on Capacity for Violence, and Survival," Boston International Security Graduate Conference, Northeastern University, February 23, 2018.

"Waiting for Al-Qaeda's next Bomb." *The Economist*, May 3, 2007. http://www.economist.com/node/9111542.

Walsh, Michael. "So Much for the 'Lone Wolf' Theory of Islamic Terrorism." *PJ Media*, February 4, 2017. https://pjmedia.com/homeland-security/2017/02/04/so-much-for-the-lone-wolf-theory-of-islamic-terrorism/.

Walsh, Nick Paton. "ISIS Defectors in Afghanistan Switch Sides." *CNN*, April 12, 2016. Middle East. http://www.cnn.com/2016/04/12/middleeast/isis-taliban-afghanistan-defectors/index.html.

Walt, Stephen M. "Revolution and War." *World Politics* 44, no. 03 (April 1992): 321–68.

Wan, Hua-Hsin. "Inoculation, Bolstering, and Combined Approaches in Crisis Communication." *Journal of Public Relations Research* 6, no. 3 (2004): 301–28.

Ward, Alex. "The War for the Capital of the ISIS 'Caliphate' Has Begun." *Vox Media*, July 6, 2017. World. https://www.vox.com/world/2017/7/6/15892630/raqqa-syria-isis-sdf-coalition-war.

Ware, B. L., and Wil A. Linkugel. "They Spoke in Defense of Themselves: On the Generic Criticism of Apologia." *Quarterly Journal of Speech* 59, no. 3 (October 1973): 273–83.

"Was the Date for the 9/11 Attack Chosen to Be the Same as the Emergency Number in the U.S. (911)?" *Skeptics*. Accessed August 6, 2017. https://skeptics.stackexchange.com/questions/10831/was-the-date-for-the-9-11-attack-chosen-to-be-the-same-as-the-emergency-number-i.

Wasmund, Klaus. "The Political Socialization of West German Terrorists." In *Political Violence and Terror: Motifs and Motivations*, edited by Peter H. Merkl, 191–228. Berkeley: University of California Press, 1986.

Wasserman, Noam, Bharat Anand, and Nitin Nohria. "When Does Leadership Matter." *Handbook of Leadership Theory and Practice* (2001): 27–63.

Watts, Clint. "ISIS and Al Qaeda Race to the Bottom." *Foreign Affairs*, November 23, 2015. https://www.foreignaffairs.com/articles/2015-11-23/isis-and-al-qaeda-race-bottom.

Watts, Clint. "When The Caliphate Crumbles: The Future of the Islamic State's Affiliates." *War on the Rocks*, June 13, 2016. https://warontherocks.com/2016/06/when-the-caliphate-crumbles-the-future-of-the-islamic-states-affiliates/.

Watts, Clint. "Why ISIS Beats Al Qaeda in Europe." *Foreign Affairs*, April 4, 2016. https://www.foreignaffairs.com/articles/2016-04-04/why-isis-beats-al-qaeda-europe.

Watts, Duncan J. *Six Degrees: The Science of a Connected Age*. New York: W. W. Norton & Company, 2004.

Weinberg, Leonard, Ami Pedahzur, and Arie Perliger. *Political Parties and Terrorist Groups*. 2nd ed. Routledge Studies in Extremism and Democracy 10. London: Routledge, 2003.

Weinstein, Jeremy M. *Inside Rebellion: The Politics of Insurgent Violence*. Cambridge Studies in Comparative Politics. Cambridge, UK: Cambridge University Press, 2006.

Weiss, Michael, and Hassan Hassan. *ISIS: Inside the Army of Terror*. New York: Regan Arts, 2015.

What Would the Fall of Mosul, Raqqa Mean for ISIS? WSJ Video, 2017. http://www.wsj.com/video/what-would-the-fall-of-mosul-raqqa-mean-for-isis/2A64093D-919B-4023-87E2-D34B37013751.html.

White, Robert. "Political Violence by the Non-Aggrieved: Explaining the Political Participation of Those with No Apparent Grievances." In *Social Movements and Violence: Participation in Underground Organizations*, edited by Donatella Della Porta, 79–103. International Social Movement Research 4. Greenwich, CT: Jai Press, 1992.

Wiktorowicz, Quintan. "A Genealogy of Radical Islam." *Studies in Conflict & Terrorism* 28, no. 2 (February 16, 2005): 75–97.

Wilkinson, Paul. "Terrorist Movements." In *Terrorism: Theory and Practice*, edited by Yonah Alexander and David Carlton. Boulder, CO: Westview Press, 1979.

Wilkinson, Paul. *Terrorism and the Liberal State*. 2nd ed. New York: New York University Press, 1986.

Williams, Carol J. "Islamic State Shows the Ability to Shift to More Sophisticated Tactics." *Los Angeles Times*, November 22, 2015. Africa. http://www.latimes.com/world/africa/la-fg-terror-tactics-20151122-story.html.

Wilson, James Q. *Bureaucracy: What Government Agencies Do and Why They Do It*. New York: Basic Books, 1989.

Wilson, Margaret Ann, and Lucy Lemanski. "Apparent Intended Lethality: Toward a Model of Intent to Harm in Terrorist Bomb Attacks." *Dynamics of Asymmetric Conflict* 6, no. 1–3 (October 29, 2013): 1–21.

Wines, Michael. "BUSH IN JAPAN; Bush Collapses at State Dinner With the Japanese." *The New York Times*, January 9, 1992. World. http://www.nytimes.com/1992/01/09/world/bush-in-japan-bush-collapses-at-state-dinner-with-the-japanese.html.

Winter, Charlie. "The Virtual Caliphate: Understanding Islamic State's Propaganda Strategy." London: Qulliam Foundation, July 2015. http://www.stratcomcoe.org/charlie-winter-virtual-caliphate-understanding-islamic-states-propaganda-strategy.

Winter, Charlie. "Trump Has Unwittingly Become an Asset for ISIS." *CNN*, November 3, 2016. Opinion. http://www.cnn.com/2016/11/03/opinions/trump-isis-both-use-fear-winter-opinion/index.html.

Winter, Charlie. "How ISIS Survives the Fall of Mosul." *The Atlantic*, July 3, 2017. https://www.theatlantic.com/international/archive/2017/07/mosul-isis-propaganda/532533/?utm_source=twb.

Wolfson, Andrew, and John Masson. "Militias Dwindle since Oklahoma City Bombing." *USA Today*, June 20, 2001. http://usatoday30.usatoday.com/news/nation/2001-04-23-mcveigh-militias1.htm.

Wood, Elizabeth Jean. "The Emotional Benefits of Insurgency in El Salvador." In *Passionate Politics: Emotions and Social Movements*, edited by Goodwin, Jeff, James M. Jasper, Francesca Polletta, and Elizabeth Jean Wood. 267–81. Chicago, IL: University of Chicago Press, 2001. doi:10.7208/chicago/9780226304007.001.0001.

Woodruff, Jody. "How Is the Islamic State Different from Other Extremist Groups?" *PBS Newshour*. Accessed October 6, 2017. http://www.pbs.org/newshour/bb/islamic-state-different-extremist-groups/.

Woods, Thomas E. *How the Catholic Church Built Western Civilization*. Washington, DC: Regency History, 2005.

Wright, Lawrence. *The Looming Tower: Al-Qaeda and the Road to 9/11*. 1st ed. New York: Alfred a Knopf Inc., 2006.

Yayla, Ahmet S. "Islamic State E-Book Released to Home-School 'Lone Wolves.'" *The Washington Times*, July 10, 2017. News.//www.washingtontimes.com/news/2017/jul/10/islamic-state-e-book-released-to-home-school-lone-/.

Yeginsu, Ceylan, and Helene Cooper. "U.S. Jets to Use Turkish Bases in War on ISIS." *The New York Times*, July 23, 2015. Europe. https://www.nytimes.com/2015/07/24/world/europe/turkey-isis-us-airstrikes-syria.html.

Zamansky, Jake. "JPMorgan Fails to Turn Its 'London Whale' Into Scapegoat." *Forbes Advisor Network*. Accessed October 6, 2017. https://www.forbes.com/sites/jakezamansky/2013/03/15/jpmorgan-fails-to-turn-its-london-whale-into-scapegoat/.

Zawodny, J. K. "Internal Organizational Problems and the Sources of Tensions of Terrorist Movements as Catalysts of Violence∗." *Studies in Conflict & Terrorism* 1, no. 3–4 (January 1978): 277–85.

Zelin, Aaron. "Free Radical." *Foreign Policy*, February 4, 2012. http://foreignpolicy.com/2012/02/04/free-radical/.

Zelkovitz, Ido. "Fatah's Embrace of Islamism." *Middle East Quarterly* 15, no. 2 (Spring 2008): 19–26.

Zenko, Micah. "10 Things You Didn't Know About Drones." *Foreign Policy*, February 27, 2012.

Zimmermann, Ekkart. *Political Violence, Crises, and Revolutions: Theories and Research*. London: Routledge, 2012. http://www.123library.org/book_details/?id=96225.

Zussman, Asaf, and Noam Zussman. "Assassinations: Evaluating the Effectiveness of an Israeli Counterterrorism Policy Using Stock Market Data." *The Journal of Economic Perspectives* 20, no. 2 (2006): 193–206.

Index

Note: "f" following a locator indicates a figure; "n" following a locator indicates an endnote; "t" following a locator indicates a table.